Anto Budiharjo

Phytostimulation by Bacillus amyloliquefaciens FZB42

Anto Budiharjo

Phytostimulation by Bacillus amyloliquefaciens FZB42

and its molecular mechanisms

Südwestdeutscher Verlag für Hochschulschriften

Impressum/Imprint (nur für Deutschland/only for Germany)
Bibliografische Information der Deutschen Nationalbibliothek: Die Deutsche Nationalbibliothek verzeichnet diese Publikation in der Deutschen Nationalbibliografie; detaillierte bibliografische Daten sind im Internet über http://dnb.d-nb.de abrufbar.
Alle in diesem Buch genannten Marken und Produktnamen unterliegen warenzeichen-, marken- oder patentrechtlichem Schutz bzw. sind Warenzeichen oder eingetragene Warenzeichen der jeweiligen Inhaber. Die Wiedergabe von Marken, Produktnamen, Gebrauchsnamen, Handelsnamen, Warenbezeichnungen u.s.w. in diesem Werk berechtigt auch ohne besondere Kennzeichnung nicht zu der Annahme, dass solche Namen im Sinne der Warenzeichen- und Markenschutzgesetzgebung als frei zu betrachten wären und daher von jedermann benutzt werden dürften.

Verlag: Südwestdeutscher Verlag für Hochschulschriften GmbH & Co. KG
Heinrich-Böcking-Str. 6-8, 66121 Saarbrücken, Deutschland
Telefon +49 681 37 20 271-1, Telefax +49 681 37 20 271-0
Email: info@svh-verlag.de

Approved by: Berlin, HU, Diss,. 2011

Herstellung in Deutschland:
Schaltungsdienst Lange o.H.G., Berlin
Books on Demand GmbH, Norderstedt
Reha GmbH, Saarbrücken
Amazon Distribution GmbH, Leipzig
ISBN: 978-3-8381-2533-6

Imprint (only for USA, GB)
Bibliographic information published by the Deutsche Nationalbibliothek: The Deutsche Nationalbibliothek lists this publication in the Deutsche Nationalbibliografie; detailed bibliographic data are available in the Internet at http://dnb.d-nb.de.
Any brand names and product names mentioned in this book are subject to trademark, brand or patent protection and are trademarks or registered trademarks of their respective holders. The use of brand names, product names, common names, trade names, product descriptions etc. even without a particular marking in this works is in no way to be construed to mean that such names may be regarded as unrestricted in respect of trademark and brand protection legislation and could thus be used by anyone.

Publisher: Südwestdeutscher Verlag für Hochschulschriften GmbH & Co. KG
Heinrich-Böcking-Str. 6-8, 66121 Saarbrücken, Germany
Phone +49 681 37 20 271-1, Fax +49 681 37 20 271-0
Email: info@svh-verlag.de

Printed in the U.S.A.
Printed in the U.K. by (see last page)
ISBN: 978-3-8381-2533-6

Copyright © 2012 by the author and Südwestdeutscher Verlag für Hochschulschriften GmbH & Co. KG and licensors
All rights reserved. Saarbrücken 2012

Summary

Bacillus amyloliqufaciens FZB42 has been known as PGPR which has an impressive effect to improve plant growth. It produces not only vast array of secondary metabolites with antibacterial and antifungal activities, but also produces the plant hormone IAA. Although many mechanisms have been elucidated, our knowledge about basic molecular mechanisms responsible for its beneficial action is far from complete. In this study, transposon mutagenesis based on *mariner* tranposon was applied to generate tranposon library which then was screened to identify the genes involved in plant growth-promoting activity. Three mutants that were impaired in their ability to colonize plant surface due to defects in biofilm formation and swarming motility were found. One mutant (*degU* mutant) showed defect in biofilm formation and swarming motility, as well, two mutants (*yusV* mutant and *pabB* mutant) impaired in biofilm formation were confirmed by complementation and retransformation. Screening by the Lemna biosystem and further assays with *A. thaliana* revealed three genes responsible for reduction in plant growth promoting activity of *B. amyloliqufaciens* FZB42. Colonization studies of these mutants in *A. thaliana* roots revealed patterns different to the wild type. A further issue pursued in this study was to discover new antibiotics using a mutant which has been blocked in its nonribosomally pathway. Screening of tranposon libraries from this mutant led to the finding of two novel ribosomally synthesized antibiotics. Further characterization revealed that these new antibiotics belonged to a novel bacteriocin (Amylocyclicin A) and a novel thiazole/oxazole-modified microcin (Plantazolicin). Last work in this study was looking for genes responsible for nematocidal production. Four mutants which showed reduction in nematocidal activity due to transposon insertion were found.

Keywords : B. amyloliquefaciens FZB42, PGPR, transposon mutagenesis, plant growth promotion

Zusammenfassung

Bacillus amyloliquenaciense FZB42 ist ein bekanntes Pflanzenwachstum-stimulierendes Rhizobakterium. Es produziert neben einer Vielzahl an Sekundärmetaboliten mit antibakterieller und antifungaler Wirkung, auch das Pflanzenhormon IAA. Obwohl viele dieser Mechanismen diskutiert werden, ist wenig darüber bekannt, auf welche Weise die Bakterien das Pflanzenwachstum fördern. In dieser Arbeit wurde eine Transposonmutagenese mithilfe des '*mariner*-transposons' durchgeführt, und so eine Transposonbibliothek erstellt. Diese wurde dann auf geeignete Phänotypen untersucht, um die Gene zu finden, welche bestimmte Phänotypen verursachen. So konnten drei Mutanten erzeugt werden, die auf Grund der gestörten Biofilmbildung und der Fähigkeit zu schwärmen die Pflanzenwurzeln nicht mehr kolonialisieren konnten. Eine solche *degU*-Mutante, welche in der Biofilmbildung und ‚Swarming' defizitär war und zwei Mutanten (*yusV* und *pabB*), die eine Beeinträchtigung in der Biofilmbildung aufwiesen, konnten durch Komplementation und Retransformation bestätigt werden. Mithilfe des Lemna-Biosystems und anderer Analysen mit *A. thaliana* konnten drei Gene bei *B. amyloliqufaciens* FZB42 gefunden werden, die wichtig für die Förderung des Pflanzenwachstums sind. Kolonisierungsexperimente der Wurzeln von *A. thaliana* mit diesen Mutanten zeigten deutlich verändertes Wachstum, verglichen mit dem Wildtypstamm. Ein weiteres Ziel dieser Arbeit war es neue Antibiotika in Mutanten, die in ihren nicht-ribosomalen Synthesen blockiert sind, zu finden. So konnten durch die Untersuchungen der Transposonbibliothek der Mutanten zwei neue Antibiotika entdeckt werden. Genauere Analysen dieser Antibiotika bestätigten, dass es sich um ein neues Bacteriocin (Amylocyclicin A) und ein neues Thiazol/Oxazole-modifiziertes Microcin

(Plantazolicin) handelt. Die abschließenden Arbeiten beschäftigten sich dann mit Untersuchungen von Genen, welche für die Produktion von Substanzen gegen Nematoden verantwortlich sind. Hierbei konnten vier Mutanten gefunden werden, die durch eine Transposoninsertion eine schlechtere.

Schlagwörter : B. amyloliquefaciens FZB42, Pflanzenwachstum-stimulierendes Rhizobakterium, Transposonmutagenese, Pflanzenwachstum förderung

Abbreviation

Amp	Ampicillin
BCIP	5-Bromo-4-chloro-3-indolylphosphat
CIAP	`Calf Intestine Alkaline Phosphatse`
CLSM	confocal laser scanning microscopy
DIG	Digoxigenin
EDTA	Ethyldiamintetraacetat
Ery	Erythromycin
EtOH	Ethanol
Fig.	Figure
h	hours
IPTG	Isopropyl β-D-thiogalactoside
Kan	Kanamycin
LB	Luria-Broth
MALDI-TOF MS	matrix-assisted laser desorption/ionization-time of flight mass spectrometry
min	minutes
MS	Murashige and Skoog
OD	optical density
ORF	open reading frame
PCR	polymerase chain reaction
rpm	rounds per minute
RT	room temperature
SEM	scanning electron microscopy
SDS	Sodiumdodecylsulfate
Spc	Spectinomycin
X-Gal	5-Bromo-4-chloro-3-indolyl-beta-D-galactopyranos

Contents

1. Introduction 13
   1.1 Plant growth-promoting rhizobacteria 13
      1.1.1 Mechanisms of plant growth-promoting rhizobacteria 14
         1.1.1.1 Direct plant growth promotion 17
         1.1.1.2 Indirect plant growth promotion 22
   1.2 Transposons mutagenesis 27
   1.3 *Bacillus amyloliquefaciens* FZB42 30
   1.4 Aims of the project 31

2. Materials and methods 32
   2.1 Chemicals and materials 32
   2.2 Plasmids, bacterial strains and primers 33
   2.3 Media and supplements 35
   2.4 Molecular biology techniques 37
      2.4.1 Standard molecular biology methods 37
      2.4.2 Transformation in *Bacillus amyloliquefaciens* FZB42 38
      2.4.3 Transposon mutagenesis 39
         2.4.3.1 Detection of *mariner* transposition events 39
         2.4.3.2 Mapping of transposon insertion sites 40
      2.4.4 Hybridization analysis of southern blots 40
         2.4.4.1 Synthesis of dig-labelled probe 40
         2.4.4.2 Preparation of samples; transfer and fixation on a membrane 40
         2.4.4.3 Hybridization and detection 41
   2.5 Screening for plant growth promotion mutants using the *lemna* biotest system 42
   2.6 Assay for plant growth promotion with *Arabidopsis thaliana* 43
      2.6.1 Sterilisation of *Arabidopsis* seeds 43
      2.6.2 Plant growth conditions 43
   2.7 Screening for biofilm, swarming and antibiotic mutants 43
      2.7.1 Screening of biofilm mutants 43

2.7.2 Screening of swarming mutants 44
2.7.3 Screening of antibiotics mutants 44

3. Results 45

3.1 Transformation *B. amyloliquefaciens* FZB42 with the transposon plasmid
TnYL B-1 45
3.2 *Himar1* transposon mutagenesis of *B. amyloliquefaciens* FZB42 47
3.3 Mapping of transposon mutagenesis 49
3.4 Discovery of genes involved in swarming motility and biofilm formation 51
    3.4.1 *B. amyloliquefaciens* FZB42 *degU*:: TnYLB-1 53
        3.4.1.1 Complementation of *degU* gene 42 54
    3.4.2 *B. amyloliquefaciens* FZB42 *yusV*:: TnYLB-1 57
        3.4.2.1 Complementation of *yusV* gene 58
    3.4.3 *B. amyloliquefaciens* FZB42 *pabB*:: TnYLB-1 61
        3.4.3.1 Complementation of *pabB* gene 61
3.5 Discovery of genes involved in plant growth-promoting activity 64
    3.5.1 *B. amyloliquefaciens* FZB42 *nfrA*:: TnYLB-1 65
        3.5.1.1 Complementation of *nfra* gene 65
        3.5.1.2 Effect of *nfrA* mutation on growth of *L. minor* and *A. thaliana* 67
    3.5.2 *B. amyloliquefaciens* FZB42 *abrB*:: TnYLB-1 70
        3.5.2.1 Complementation of *abrB* mutant 70
        3.5.2.2 Effect of *abrB* mutation on growth of *L. minor* and *A. thaliana* 72
    3.5.3 *B. amyloliquefaciens* FZB42 *RBAM_017410*:: TnYLB-1 75
        3.5.3.1 Complementation of *RBAM_017410* mutant 76
        3.5.3.2 Effect of *RBAM_017410* mutation on growth of *L. minor* and
A. thaliana 77
3.6 Colonization of *B. amyloliquefaciens* FZB42 and its mutants in *A. thaliana* roots
growing in gnotobiotic system 81
3.7 MALDI-TOF MS analysis of metabolite released by *B. amyloliquefaciens* FZB42
in plant-bacteria interactions... 89
3.8 Screening of antibiotic mutants 91

3.9 Screening of nematocidal mutants 93

4. Discussion **94**

    4.1 *Himar1* transposon mutagenesis of *B. amyloliquefaciens* FZB42 95

    4.2 Identification of genes involved in swarming motility and biofilm formation in

        B. amyloliquefaciens FZB42 genome.. 97

    4.3 Identification of genes involved in plant growth promoting activity and

        colonization of the mutants in the root of *A. thaliana* 104

    4 4. Identification of genes involved in production of antibiotic and

        Nematocidal 110

5. Reference **112**

List of Figures

Figure 1.	Illustration of the most important mechanisms of biological control of plant diseases by bacteria	23
Figure 2.	Restriction analysis of plasmid DNA cut with *EcoRI*	47
Figure 3.	TnYLB-1 transposition in *B. amyloliqufaciens* FZB42	49
Figure 4.	Distribution of randomly TnYLB-1 insertions in the *B. amyloliquefaciens* FZB42 chromosome	51
Figure 5.	Genomic organization of *degU* region carrying the TnYLB-1 insertion and its flanking regions	53
Figure 6.	Strategy for construction of pUC18-ΔdegU cassette	55
Figure 7.	PCR product of *degU* gene	56
Figure 8.	Phenotype of swarming motility in *degU* mutant	56
Figure 9.	Phenotype of biofilm formation in *degU* mutant	57
Figure 10.	Genomic organization of *yusV* region carrying the TnYLB-1 insertion and its flanking regions	58
Figure 11.	Strategy for construction of pUC18-Δ*yusV* cassette	59
Figure 12.	PCR product of *yusV* gene	60
Figure 13.	Phenotype of biofilm formation in *yusV* mutant	60
Figure 14.	Genomic organization of *pabB* region carrying the TnYLB-1 insertion and its flanking regions	61
Figure 15.	Strategy for construction of pUC18-Δ*pabB* cassette	62
Figure 16.	PCR product of *pabB* gene	63
Figure 17.	Phenotype of biofilm formationt in *pabB* mutant	64
Figure 18.	Genomic organization of *nfrA* region carrying the TnYLB-1 insertion and its flanking regions	65
Figure 19.	Strategy for construction of pUC18-ΔnfrA cassette	66
Figure 20.	PCR product of *nfrA* gene	67
Figure 21.	Influence of *nfrA* mutation on plant growth-promoting ability of *B. amyloliquefaciens* FZB42 on *L. minor*	67
Figure 22.	Growth stimulating effects of *nfrA* mutation on *L. minor*	68

Figure 23.	Influence of *nfrA* mutation on plant growth-promoting ability of *B. amyloliquefaciens* FZB42 on *A. thaliana*	69
Figure 24.	Growth stimulating effects of *nfrA* mutation on *A. thaliana*	69
Figure 25.	Genomic organization of *abrB* region carrying the TnYLB-1 insertion and its flanking regions	70
Figure 26.	Strategy for construction of pUC18-ΔabrB cassette	71
Figure 27.	PCR product of *abrB* gene	72
Figure 28.	Influence of *abrB* mutation on plant growth promoting ability *B. amyloliquefaciens* FZB42 on *L. minor*	73
Figure 29.	Growth stimulating effects of *abrB* mutation on *L. minor*	73
Figure 30.	Influence of *abrB* mutation on plant growth promoting ability of *B. amyloliquefaciens* FZB42 on *A. thaliana*	74
Figure 31.	Growth stimulating effects of the *abrB* mutation on *A. thaliana*	75
Figure 32.	Genomic organization of *RBAM_017410* region carrying the TnYLB-1 insertion and its flanking regions	75
Figure 33.	Strategy for construction of pUC18-Δ *RBAM_017410* cassette	76
Figure 34.	PCR product of *RBAM_017410* gene	77
Figure 35.	Influence of *RBAM_017410* mutation on plant growth promoting ability *B. amyloliquefaciens* FZB42 on *L. minor*	78
Figure 36.	Growth stimulating effects of *RBAM_017410* mutation on *L. minor*	79
Figure 37.	Influence of RBAM-017410 mutation on plant growth promoting ability of *B. amyloliquefaciens* FZB42 on *A. thaliana*	80
Figure 38.	Growth stimulating effects of *RBAM_017410* mutation on *A. thaliana*	80
Figure 39.	CLSM Image of *B. amyloliquefaciens* FZB42 on *Arabidopsis* roots	82
Figure 40.	SEM of *B. amyloliquefaciens* FZB42 colonizing *A. thaliana* roots	82
Figure 41.	CLSM Image of *yusV* mutant on *A. thaliana* roots	83
Figure 42.	SEM of *yusV* mutant colonizing *A. thaliana* roots	84
Figure 43.	CLSM Image of *degU* mutant on *A. thaliana* roots	84
Figure 44.	SEM of *degU* mutant colonizing *A. thaliana* roots	85
Figure 45.	CLSM Image of *nrfA* mutant on *A. thaliana* roots	85
Figure 46.	SEM of *nfrA* mutant colonizing *A. thaliana* roots	86

Figure 47.	CLSM Image of *abrB* mutant on *A. thaliana* roots	86
Figure 48.	SEM of *abrB* mutant colonizing *A. thaliana* roots	87
Figure 49.	CLSM Image of *RBAM_017410* mutant on *A. thaliana* roots	87
Figure 50.	SEM of *RBAM_017410* mutant colonizing *A. thaliana* roots	88
Figure 51.	MALDI-TOF MS analysis of surfactin produced by *B. amyloliquefaciens* FZB42 and its mutants	91
Figure 52.	Spot on lawn test of WY01 mutant	93

List of Tables

Table 1.	Chemicals and materials used in the present study	32
Table 2.	Plasmids used in the present study	33
Table 3.	Bacterial strains used in the present study	34
Table 4.	Primers used in this study	34
Table 5.	Supplements	36
Table 6.	Average of transposon frequency	48
Table 7.	*In vitro* effects of *B. amyloliquefaciens* FZB42 and its mutants on *C. elegans* livability	97

1. Introduction

1.1 Plant growth-promoting rhizobacteria

In the rhizosphere, that is the portion of soil on the plant root or its close vicinity, bacteria are abundantly present, most often organized in microcolonies (Bloemberg et al. 2001). The plant rhizosphere is an essential soil ecological environment for plant–microorganism interactions, which include colonization by a variety of microorganisms in and around the roots that may result in symbiotic, endophytic, associative, or parasitic relationships within the plant, depending on the type of microorganisms, soil nutrient status, and soil environment (Albareda et al. 2006). In this sphere, intensive interactions are taking place between the plant, soil, soil microfauna and microorganisms, where bacteria are the most abundant microorganisms (Antoun and Kloepper, 2001). The region around the root is relatively rich in nutrients because as much as 40% of plant photosynthates are lost from the roots, hence it supports the large microbial population (Ping et al. 2004).

The activity and diversity of microorganisms adjacent to roots differs from the activity and diversity of the microorganisms in the bulk soil (Wang et al. 2005). In the bulk soil population sizes were larger, but in the rhizosphere the phylogenetic diversity is more restricted (Marilley and Aragno, 1999; Berg et al. 2005). The high concentration of easily metabolizable organic compounds in the rhizosphere sustain microbial populations that are more active, denser but less diverse than those present in bulk soil (Inbar et al. 2005).

Rhizobacteria are rhiszosphere competent bacteria that colonize and proliferate on all the ecological niches found on the plant roots at all stages of plant growth, in the presence of a competing microflora (Antoun and Kloepper, 2001). Based on their effects on the plant, microbes interacting with plants can be categorized as pathogenic, saprophytic and beneficial. Pathogens can attack leaves, stems or roots. Microbes in their interactions with plants, no matter whether the microbe is beneficial or pathogenic, often use the same mechanisms, although in different combinations and for different purposes. Similarly, it is obvious that microbes in their interaction with plants use similar strategies as in their interactions with other eukaryotes such as fungi and humans (Lugtenberg *et al.* 2002). Some of these rhizobacteria not only benefit from the nutrients secreted by the plant root but also beneficially influence the plant in a direct or indirect way, resulting in a stimulation of its growth (Bloemberg *et al.* 2001).

Plant growth-promoting rhizobacteria (PGPR), first defined by Joseph W. Kloepper and Milton N. Schroth, include a wide range soil bacteria that colonize the roots of plants following inoculation onto seed and enhance plant growth by increasing seed emergence, plant weight, and crop yields (Ping *et al.* 2004 and Ryu *et al.* 2004). Besides colonizing the root surfaces and the closely adhering soil interface PGPR can also enter root interior and establish endophytic populations. Many of them are able to transcend the endodermis barrier, crossing from the root cortex to the vascular system, and subsequently thrive as endophytes in stem, leaves, tubers, and other organs. The extent of endophytic colonization of host plant organs and tissues reflects the ability of bacteria to selectively adapt to these specific ecological niches. Consequently, intimate associations between bacteria and host plants can be formed without harming the plant. Although, it is generally assumed that many bacterial endophyte communities are the product of a colonizing process initiated in the root zone, they may also originate from other source than the rhizosphere, such as the phyllosphere, the anthosphere, or the spermosphere (Compant *et al.* 2005).

PGPRs have drawn much attention in recent years because of their contribution to the biological control of plant pathogens and the improvement of plant growth. Inoculation of plants with 'dual' microbial inoculants, or even a consortium of them, is becoming more important in a framework of sustainable agriculture for the advantage their beneficial effects afford, providing there is no competition between inoculants (Albareda *et al.* 2006). Extensive research has demonstrated that PGPRs could have an important role in agriculture and horticulture in improving crop productivity. In addition, these organisms are also useful in forestry and environmental restoration. As agricultural production intensified over the past few decades, producers became more and more dependent on agrochemicals as a relatively reliable method of crop protection helping with economic stability of their operations. However, increasing use of chemical inputs causes several negative effects, i.e., development of pathogen resistance to the applied agents and their nontarget environmental impacts. Furthermore, the growing cost of pesticides, particularly in less-affluent regions of the world, and consumer demand for pesticide-free food has led to a search for substitutes for these products. There are also a number of fastidious diseases for which chemical solutions are few, ineffective, or nonexistent. PGPR is thus being considered as an alternative or a supplemental way of reducing to the use of chemicals in agriculture in many different applications (Lucy *et al.* 2004 and Compant *et al.* 2005).

1.1.1 Mechanisms of plant growth-promoting rhizobacteria

Diverse mechanisms are involved in plant-bacteria interactions, and in many cases individual PGPR have several mechanisms on their activities to promote the plant growth at various times during the life cycle of the plant (Glick *et al.* 1999; Berg *et al.* 2002 and Müller, 2009). Based on their mode of action, PGPRs are grouped into two large classes, namely the PGPRs that directly affect plant metabolism resulting in increased plant growth, seed emergence or improved crop yields and the Biocontrol-PGPRs, which suppress plant pathogens, thereby benefiting the plant indirectly (Ping and Boland, 2004; and Wang, 2005).

In all mode of action of PGPR, the ability to colonize plant habitats especially roots is important for all successful plant–microbe interactions, which in turn determine inoculum efficacy both for crop yield enhancement and for disease control. This has led to an emphasis on selection of plant-beneficial bacteria that are rhizosphere competent (i.e., beneficial bacteria that effectively colonize the root system) (Kamilova *et al.* 2005 and Compant *et al.* 2005). Steps of colonization include recognition, adherence, invasion (only endophytes and pathogens), colonization and growth, and several strategies to establish interactions. Plant roots begin crosstalk with soil microbes by generating signals that are recognized by the microbes, which in turn produce signals that initiate colonization (Bais *et al.* 2006). In *Pseudomonas fluorescens* WCS365 the major traits involved in competitive root tip colonization are motility; adhesion to the root; a high growth rate in root exudate; synthesis of amino acids, uracil, and vitamin B1; the presence of the O-antigenic side chain of lipopolysaccharide; the two-component ColR/ColS sensory system; fine-tuning of the putrescine uptake system (the mutant had an impaired *pot* operon); the site-specific recombinase Sss or XerC; the *nuo* operon (the mutant had a defective NADH:ubiquinone oxidoreductase); the *secB* gene involved in a protein secretion pathway; and the type three secretion system (TTSS) (Lugtenberg *et al.* 2001).

1.1.1.1 Direct plant growth promotion

Even though the molecular basis for the interactions is not always well known, several basic principles of molecular interplay between the PGPRs and plants have been successfully unraveled. The most prominent example is nitrogen fixation by bacteria such as *Rhizobium* and *Bradyrhizobium* that can form nodules on roots of leguminous plants such as soybean, pea, peanut, and alfalfa, and convert N_2 into ammonia, which in contrast to N_2 can be used by the plant as a nitrogen source. The symbiosis between rhizobia and its legume host plants is an important example for plant growth-promoting rhizobacteria (PGPR). The symbiosis is initiated by the formation of root or stem nodules in response to the presence of the bacterium. Lipooligosacharide signal molecules that are secreted by the bacterium play a crucial role in this process. The bacteria penetrate the cortex, induce root nodules, multiply and subsequently differentiate into bacteroids, which produce the nitrogenase enzyme complex. Within the root nodules, the plant creates a low oxygen concentration, which allows bacterial nitrogenase to convert atmospheric nitrogen into ammonia. In return, the plant supplies the bacteria with a carbon source. The molecular interaction between the plants (providing the carbon source) and the microorganisms (providing the nitrogen supply) is highly complex and involves many factors (Freiberg *et al.* 1997; Bloemberg and Lugtenberg, 2001; Berg, 2009; and

Lugtenberg and Kamilova, 2009). However, several bacteria belonging to the genus *Azospirillum*, *Burkholderia*, and *Stenotrophomonas* have the ability to fix nitrogen as a free living organism (Dobbelare *et al.* 2003).

Some bacteria are able to influence the hormonal balance in the plant. For example, *Pseudomonas putida* GR12-12 and *Enterobacter cloacae* UW4 contain the gene for ACC (1-aminocyclopropane-1-carboxylate) deaminase, which can cleave the plant ethylene precursor ACC, and thereby lower the level of ethylene in a developing or stressed plant (Hall *et al.* 1996 and Hontzeas *et al.* 2004). PGPR that contain the enzyme ACC deaminase, when bound to the seed coat of a developing seedling, provides a mechanism for ensuring that the ethylene level does not become elevated to the point where root growth is impaired. With the longer roots, survival of some seedlings will be enhanced especially during the first few days after the seeds are planted. Similarly, ACC deaminase-containing bacteria bound to the roots of plants can act as a sink for ACC and protect stressed plants from some of the deleterious effects of stress ethylene (Glick, 2005). Several other forms of stress are relieved by ACC deaminase producers, for example effects of phytopathogenic bacteria, and resistance to stress from polyaromatic hydrocarbons, from heavy metals such as Ca^{2+} and Ni^{2+}, and from salt and draught (Glick *et al.* 2007).

Mineral supply is also involved in plant growth promotion and low levels of soluble phosphate can limit the growth of plants. Some plant-growth promoting bacteria solubilize insoluble phosphate from either organic or inorganic bound phosphates which makes phosphorous available to the plants (Rodriguez and Fraga, 1999). *Pseudomonas fluorescens* NJ-101, *Pseudomonas fluorescens* EM85 and *Bacillus amyloliquefaciens*

FZB45 are to name of some bacteria that have ability to solubilize insoluble phosphate, therefore enhance nutrient availability to plants and facilitating plant growth (Idris *et al.* 2002; Bano and Mussarat, 2004; Dey *et al.* 2004 and Vassilev *et al.* 2006). Idris *et al.* concluded that phytase activity of *B. amyloliquefaciens* FZB45 is important for plant growth stimulation under phosphate limitation. Extracellular pyhtase activity is mainly produced during the late stage of exponential growth and during the transition to stationary growth phase, suggesting that similar to other extracellular depolymerases phytase acts as a `scavenger' enzyme after exhaustion of rapidly metabolized nutrient sources (Idris *et al.* 2002). Another mineral that is important for the plant growth is iron. The shortage of bioavailable iron in soil habitats and on plant surfaces generates an intense competition among microorganisms. By far, the most common mechanism of iron acquisition by microorganisms involves chelation of ferric iron by siderophores. Under iron-limiting conditions PGPR produce low-molecular-weight compounds called siderophores to competitively acquire ferric ion. The release of siderophores chelates iron and makes it available to the plant root (Loper and Henkels, 1997; Ping and Boland, 2004; and Katiyar and Goel, 2004)

Phytohormones are involved in the control of growth and in almost every important developmental process in plants. Many PGPR can produce phytohormones, such as auxins, cytokinins, and gibberellins (Salamone *et al.* 2001; Ortiz-Castro *et al.* 2008; and Joo *et al.* 2009). Indeed, three types of plant growth promoting substances have been detected in the supernatant of *Azospirillum* cultures, these are auxins, cytokinins and gibberellines. Of these, the auxin IAA (indole-3-acetic acid) is quantitatively the most important one as it can directly benefit the plant root system by promoting the development of lateral roots and apical meristem divisions that lead to lengthening of the roots (Malhotra and Srivastava, 2009). Experiments with *Azospirillum* mutants altered in IAA production prove the view that after *Azospirillum* inoculation IAA causes increased rooting, which in turn enhances mineral uptake (Steenhoudt and Vanderleyden, 2000). Idris *et al.* (2007) proved that biosynthesis of IAA in *B. amyloliquefaciens FZB42* affected its ability to promote plant growth. By inactivating the genes involved in tryptophan biosynthesis and in a putative tryptophan-dependent IAA biosynthesis pathway, IAA concentration and plant growth promoting activity in respective mutants were reduced. Gibberellins are plant hormones which control several different physiological processes such as the stimulation of stem elongation by stimulating cell division and elongation, the stimulation of bolting/flowering in response to long days, the break of seed dormancy in some plantswhich require stratification or light to induce germination, the stimulation of enzyme production (α-amylase) in germinating cereal grains for mobilization of seed reserves, the induction of maleness in dioecious flowers (sex expression), the inducement of a parthenocarpic (seedless) fruit development, and the retardation of senescence in leaves and citrus fruits (Joo *et al.* 2004). Production of gibberellins which promote plant growth has been reported in different bacteria such as *Rhizobium Phaseoli, Acetobacter diazotrophicus, Herbaspirillum seropedicae, B. pumilus, B. licheniformis* and B. macroides (Joo *et al.* 2005; Atzhorn *et al.*1988; Bastian *et al.* 1998; and Gutierrez-Manero *et al.* 2001). Cytokinins are a class of phytohormones produced by plants and microorganisms which may play an essential role in regulating cytokinesis, growth and development in plants (Aloni *et al.* 2006). Hence, it can be expected that plant inoculation with PGPR capable of producing cytokinins may increase the level of cytokinins in root tissues which in turn may have an impact on plant growth (Ortis-Castro et al. 2008). Cytokinins are thought to be the signals involved in mediating of environmental stress from roots to shoots (Jackson, 1993).

Some rhizobacteria, such as strains from *B. subtilis*, *B. amyloliquefaciens,* and *Enterobacter cloacae*, promote plant growth and induce systemic resistance by releasing volatile organic compound (VOC) (Ryu et al. 2003; and Ryu et al. 2004). Analysis of the volatiles emitted from *Bacillus subtilis* GB03 and *Bacillus amyloliquefaciens* IN937a, revealed that two compounds, 3-hydroxy-2-butanone (acetoin) and 2,3-butanediol, were shared by both bacterial strains whereas other PGPR strains that did not trigger enhanced growth via volatile emissions also did not share this same subset of volatile components. Other components of the complex bouquet from *B. subtilis* (e.g. decane, undecane, undecane-2-one, tridecan-2-one and tridecan-2-ol) were not active. Furthermore, pharmacological applications of 2,3-butanediol enhanced plant growth whereas bacterial mutants blocked in 2,3-butanediol and acetoin synthesis were inactive in plant growth promotion (Ryu et al. 2003). VOC can also trigger the growth of the plant by regulating auxin homeostatis in which the gene expression for auxin production was upregulated. In addition, microarray data revealed coordinated regulation of cell wall loosening enzymes that implicated cell expansion with *B. subtilis* GB03 exposure (Zang et al. 2007).

The cofactor Pyrroloquinoline quinone (PQQ) is recently regarded as a plant growth promotion factor produced by *Pseudomonas fluorescens* B16 (Choi et al. 2007). In mammals, pyrroloquinoline quinone (PQQ) functions as a potent growth factor, although its biological functions are not fully understood (Steinberg et al. 1994).

Mutations in *pqq* genes abolished plant growth-promotion activity of wild-type B16, whereas synthetic PQQ promotes growth of tomato and cucumber plants. This study provides evidence that PQQ is a plant growth-promotion factor because of its antioxidant activity. However, it cannot be excluded that the effect is indirect because PQQ is a cofactor of several enzymes, e.g., involved in antifungal activity and induction of systemic resistance (Choi *et al.* 2007).

1.1.1.2 Indirect plant growth promotion

Pathogenic microorganisms which damage the plant health are a major and chronic threat to food production and ecosystem stability worldwide. Over the past few decades, producers became more dependent on agrochemicals as a relatively reliable method of crop protection in order to intensify their agricultural production. However, increasing use of chemical inputs causes several negative effects, i.e., development of pathogen resistance to the applied agents and their nontarget environmental impacts as well as negative impact to human health (Gerhadson, 2002; and Leach and Mumford, 2008). The use of microbes as form of biological control to manage diseases is an environment-friendly approach. These biocontrol agents are a natural enemy of the pathogen, and if they produce secondary metabolites, they do only locally, on or near the plant surface, i.e., the site where they should act (Lugtenberg and Kamilova, 2009). Such microorganisms can produce substances that may limit the damage caused by phytopathogens, e.g. by producing antibiotics, siderophores, and a variety of enzymes and can also function as competitor of pathogens for colonization of sites and nutrients (Timmusk, 2003). Some PGPR strain can also lead to a state of induced systemic resistance (ISR) in the treated plant. ISR occurs when the plant's defense mechanisms are triggered and primed to resist infection by pathogens (Van Loon, 1998). Schematic illustration of some important mechanism of biological control of plant diseases by bacteria is shown in Fig. 1.

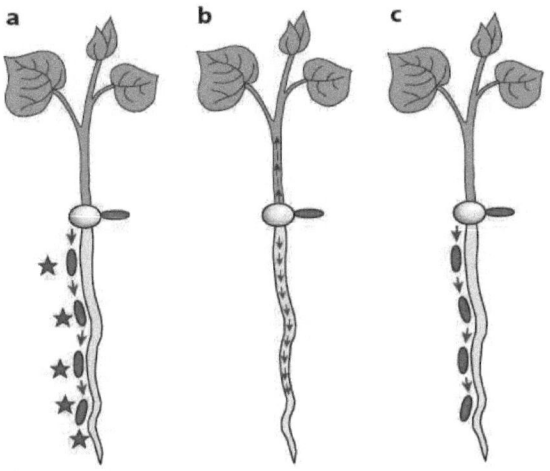

Figure 1. Illustration of the most important mechanisms of biological control of plant diseases by bacteria. In all cases illustrated here, biocontrol begins by coating seeds with the biocontrol bacterium. (a) Antibiosis. The bacterium colonizes the growing root system and delivers antibiotic molecules around the root, thereby harming pathogens that approach the root (indicated by stars). (b) Induced systemic resistance (ISR). Local root colonization is sufficient to induce ISR. Many bacterial products induce systemic signaling, which can result in protection of the whole plant against diseases caused by different organisms. The latter aspect of ISR resembles innate immunity in humans and animals. (c) Competition for nutrients and niches. Biocontrol bacteria acting through this mechanism excel in fast chemotactic movement along the growing root in their efficient hunt for root exudate components, thereby outcompeting the pathogen in scavenging nutrients and in occupying niches on the root (Lugtenberg and Kamilova 2009).

PGPR can produce a variety of antibiotics including 2,4-diacetylphloroglucinol (DAPG), phenazines, hydrogen cyanide, pyrrolnitrin, pyoluteorin, viscosinamide and tensin produced by pseudomonads (Nielsen *et al.* 1999; Nielsen *et al.* 2000; Bloemberg and Lugtenberg, 2001; and Haas and Defago, 2005) and zwittermycin A, kanosamine, bacillomycin D and fengycin produced by *Bacillus* (Raaijmakers *et al.* 2002; and

Koumoutsi *et al.* 2004). Biocontrol agents from *P. fluorescens* act rather nonspecific in their ability to protect plants from soil phytopathogens. Indeed, each strain can typically work in more than one pathosystem, i.e. protect more than one plant species from often distinct pathogens, provided the rhizosphere is successfully colonized. They have been mostly studied for protection of crop plants from phytopathogenic oomycetes and fungi and to a lesser extent bacteria and nematodes (Couillerot *et al.* 2009). The production of anti-fungal metabolites (AFMs) in *Pseudomonas* involves a complex regulation. Main factors in the regulation of the biosynthesis of most AFMs are global regulation and quorum sensing. Global regulation is directed by the *gacS/gacA* genes, which encode a two-component regulatory system that senses an as yet unknown signal(s). Quorum sensing involves the production of *N*-acyl homoserine lactone (AHL) signal molecules by an AHL synthase such as LuxI. The AHL then binds to and activates a transcriptional regulator, such as LuxR. The activated form of the transcriptional regulator then stimulates gene expression (Bloemberg and Lugtenberg, 2001). An antibiotic produced by *Bacillus cereus* and *Bacillus thuringiensis*, zwittermycin A, adversely affects the growth and activity of a wide range of microorganisms, including several plant pathogenic fungi and in particular *Phytophthora* and *Pythium* species (Raaijmakers *et al.* 2002).

Various PGPR can reduce the activity of pathogenic microorganisms not only through microbial antagonisms, but also by inducing a state of systemic resistance in plants, which provides protection against a broad spectrum of phytopathogenic organisms including fungi, bacteria and viruses. This enhanced defensive capacity is termed induced systemic resistance (ISR) (Van loon, 2007). The mechanisms of ISR include (1) developmental—escape: linked to growth promotion, (2) physiological—tolerance: reduced symptom expression, (3) environmental: associated with microbial antagonisms in the rhizosphere, and (4) biochemical—resistance: induction of cell wall reinforcement, induction of phytoalexins and pathogenesis-related proteins, and "priming" of defense responses (Berg, 2009). The evidence of the ISR was first described by Van Peer et al. (1991) in carnation that was systemically protected against *Fusarium oxysporum* f.sp. *dianthi* upon treatment with strain *Pseudomonas fluorescens* WCS417 and by Wei et al. 2001 on cucumber (*Cucumis sativus*) with reduced susceptibility to foliar disease caused by *Colletotrichum orbiculare*. Before challenge inoculation, no increase in phytoalexin levels could be detected in induced and uninduced plants but, upon subsequent inoculation with *F. oxysporum* f.sp. *dianthi*, phytoalexin levels in ISR-expressing plants rose significantly faster than in uninduced plants. *Bacillus pumilus* SE34 induces ISR in bean (*Phaseolus vulgaris*) against the root-rot fungus *F. oxysporum* f.sp. *pisi.* by appositions containing large amounts of callose and phenolic materials, thereby effectively preventing fungal ingress (Benhamou et al. 1996). Studies on mechanisms show that elicitation of ISR in *Bacillus* spp is associated with ultrastructural changes in plants during pathogen attack and with cytochemical alterations (Kloepper et al. 2004). ISR acts through a different signaling pathway to that regulating systemic acquired resistance (SAR), the ISR pathway is induced when the plant is challenged by non-pathogenic organism. Bacterial determinants that are responsible to trigger ISRs include siderophores, the O-antigen of lipopolysacharide, N-acyl-homoserine lactones, salicylic acid and VOCs (e.g., 2,3-butandiol). Whereas some PGPR activate defense-related gene expression, other examples appear to act solely through priming of effective resistance mechanisms, as reflected by earlier and stronger defense reaction once infection occurs (Bloemberg and Lugtenberg 2001; Conrath et al. 2002; and Berg, 2009). Investigations of the signal transduction pathways of elicited plants suggest that *Bacillus* spp. activate some of the same pathways as *Pseudomonas* spp. Pseudomonad PGPR that trigger ISR is dependent on JA, ethylene, and *Npr1*, a regulatory gene that encodes salicylate dehydrogenase, but independent of SA, a result that is in agreement with several strain of *Bacillus* spp. However, in other cases, ISR elicited by *Bacillus* spp. is dependent on salicylic acid and independent of jasmonic acid and *NPR1*. The VOCs of *Bacillus subtilis* GB03 and *Bacillus amyloliquefaciens* IN937a that trigger ISR involved signal transduction pathways that were independent of SA, JA, and *Npr1*. In addition, in some cases ISR by *Bacillus* spp leads to accumulation of the defense gene *PR1* in plants, ISR by *Pseudomonas* spp. does not. (Kloepper *et al.* 2004; and Ryu *et al.* 2004).

Competition for niche and nutrients can also be a fundamental mechanism by which PGPB protect plants from phytopathogens. In the rhizosphere there are various suitable nutrient-rich niches as a result of exudation of compounds attracting a great diversity of microorganisms, including phytopathogens (Compant *et al.* 2005). Known chemical attractants present in root exudates include organic acids, amino acids, and specific sugars (Welbaun *et al.* 2004). Some exudates can also be effective as antimicrobial agents and thus give ecological niche advantage to organisms that have adequate enzymatic machinery to detoxify them. This implies that PGPR competence highly depends either on their abilities to take advantage of a specific environment or on their abilities to adapt to changing conditions (Bais *et al.* 2004). Competition may concern the acquisition of organic substrates released by seeds and roots as well as micronutrients such as soluble iron, which is often in limiting amounts in soil. Iron acquisition entails the production of iron transporters (siderophores), noticeably fluorescent pyoverdines (Couillerot *et al.* 2009). Although various bacterial siderophores differ in their abilities to sequester iron, in general, they deprive pathogenic fungi of this essential element since the fungal siderophores have lower affinity (Loper *et al.* 1999).

1.2 Transposons mutagenesis

Transposons are mobile genetic elements that can move from one site to another in the genome with the aid of a recombinase called a transposase. They are ubiquitous and present in Eubacteria, Archaea, and Eukarya, including in humans in which they constitute a significant fraction of the genome. Transposons are widely used as tools for random mutagenesis in vitro and in vivo in a variety of organisms ranging from gram-negative Escherichia coli to eukaryotes, and engineered transposons have been developed that incorporate a variety of useful features (Bordi *et al.* 2008; and Petzke and Luzhetskyy, 2009). Transposable elements are the causative agents of various insertion, deletion, inversion and chromosomal fusion mutations. When inserted in the appropriate location of the genome, mutation caused by transposons can inactivate or activate critical genes (Chandler and Mahillon, 2002; and Reznikoff, 2003). Transposable elements in bacteria range from simple insertion sequence (IS) elements that consist of a gene(s) for transposition bounded by inverted repeat sequences, to composite transposons composed of a pair of IS elements that bracket additional genetic information for antibiotic resistance or other properties, to more complex conjugative transposons that exhibit hybrid properties of transposons, plasmids, and bacteriophages (Hayes, 2003). Numerous transposon delivery systems have been developed for *Escherichia coli* and other gram negative bacteria. However, in many cases these incorporate selectable markers that are not conducive to their use in gram-positive bacteria (Bordi *et al.* 2008). There are two mechanisms in transposon movement, namely 'cut and paste' and 'replicative transposition'. In cut and paste mechanism the element is excised from its resident location and inserted at a new position, whereas in replicative transposition, the transposition process involves cointegration of the donor replicon that harbors the transposon and the target molecule with concomitant duplication of the transposon (Hayes, 2003).

The transposons most favored as genetic tools are those that insert randomly or near-randomly, or can be manipulated to behave in this way. Tn*917* transposon, a streptococcal Tn*3*-like transposon, was the first transposon developed for use in *B. subtilis*. It was adapted by the incorporation of a promoterless *lacZ* gene, and the resulting Tn*917lac* transposon was used to generate large numbers of reporter fusions. Despite their wide use, Tn*917* has significant shortfalls in which ninety-nine percent of all Tn*917* insertions occur at several "hot-spot" regions of the *B. subtilis* chromosome (Youngman *et al.* 1983; and Youngmann *et al.* 1985). More recently, Tn*10* and *mariner* transposon were used for in vivo transposition in *B. subtilis*. Tn*10*, a transposon isolated from *E. coli*, was adapted for *B. subtilis* by fusion of the transposase gene to expression signals appropriate for this bacterium (Petit *et al.* 1990). Unlike Tn*917*, Tn*10* does not appear to have preferred insertion sites in the *B. subtilis* chromosome; but it is known to have a strong preference for a 6-bp target sequence. Hence reduces the number of potential Tn*10* insertion sites on the *B. subtilis* chromosome and, as a consequence,

Tn*10*'s effectiveness as a tool for random mutagenesis (Halling and Kleckner, 1982; and Breton *et al.* 2006).

Mariner-family transposable elements are a diverse and taxonomically widespread group of transposons occurring throughout the animal kingdom and especially prevalent in insects. Their wide distribution results from their ability to be disseminated among hosts by horizontal transmission and also by their ability to persist in genomes through multiple speciation events (Robertson, 1993; and Hartl *et al.* 1997). Among hundreds of different mariners that have been detected, only two are known to be active. The first is *Mos1* which was discovered from *Drosophila mauritiana*. The second is the *Himar1* which was isolated from the horn fly *Haematobia irritans*. A transposon based on the eukaryotic *mariner* family of transposons has been used for eubacteria, archaebacteria, and eukaryotic cells (Lampe *et al.* 1999; and Julian and Fehd, 2003). Compared to other transposons that have been engineered to construct insertional mutagenesis in bacteria, *mariner* elements offer several advantages. First, they do not require species-specific host factors for efficient transposition. Second, apart from the dinucleotide TA, mariner elements have no specific sequence requirements for their insertions. Third, they transpose in both eukaryotes and prokaryotes. In addition, transformation with mariner elements usually leads to 10-fold-more mutants than transformation with the Tn*917* (Louvel *et al.* 2005; and Picardeau, 2010). Due to its effectiveness in transposition, numerous transposon systems based on the mariner transposon family have been applied for mutagenesis in bacteria (Bourhy *et al.* 2005; Wu *et al.* 2006; Liu *et al.* 2007; and Kritisch *et al.* 2008).

1.3 *Bacillus amyloliquefaciens* FZB42

Among various group of plant-associated microorganisms, strains of *Bacillus* have gained more attention as they have several advantages over other biocontrol bacteria in that they are easy to cultivate and store. In addition, they offer a biological solution to the formulation problem due to their ability to form heat- and desiccation-resistant spores, which can be formulated readily into stable products. Hence they can be applied as spores on plant seeds or in inoculants (Reva *et al.*, 2004; and Emmert and Handelsmann, 1999). The genus Bacillus is characterised by gram positive, rod shaped, facultative aerobe, endospore forming bacteria that live in soil and often colonise the plant rhizosphere. It has a broad host range, ability to produce different kind of antibiotics and other secondary metabolites important for plant growth (Gardener, 2004).

B. amyloliquefaciens FZB42 is regarded as PGPR due to its biocontrol and phytostimulator activity. Its genome has been sequenced and mapped; therefore it is possible to detect the genes responsible for its plant growth activity (Chen *et al.* 2007). Phytase activity and auxin production of *B. amyloliquefaciens* FZB42 which are important for plant growth promotion have been reported (Idriss *et al.*, 2002; and Idriss *et al.* 2007). FZB42 genome analysis revealed the presence of numerous gene clusters involved in synthesis of non-ribosomally synthesized cyclic lipopeptides and polyketides with distinguished antimicrobial action (Chen *et al.* 2009a; Chen *et al.* 2009b). For example production of non-ribosomally synthesized peptides such as *bacillomycin D* and *fengycin* are able to inhibit growth of phytopathogenic fungi such as *Fusarium oxysporum* in synergistic way (Koumoutsi *et al.*, 2004). Whereas polyketide compounds such as difficidin and bacilysin act efficiently against fire blight disease caused by *Erwinia amylovora* (Chen *et al.* 2009c).

1.4 Aims of the project

B. amyloliquefaciens FZB42 is known as plant growth promoting bacterium due to production of a vast array of secondary metabolites which protect and support the growth of the plant. Several mechanisms of its activities have been reported recently (Idriss, *et al.* 2007; Koumoutsi *et al.*, 2004; and Chen *et al.* 2009c), however, still many mechanisms are not fully understood.

The complete genome sequence of *B. amyloliquefaciens* FZB42 showed that many regions in this genome were still obscure (Chen *et al*, 2007); hence it needs to be exploited further in order to reveal the unexpected potential for developing agrobiotechnological agents with predictable features. In doing so, transposon mutagenesis will be applied to discover genes that are potentially involved in its plant growth-promoting activity. The *mariner*-based transposon TnYLB-1 was selected, since it "jumps" into the *B. subtilis* chromosome with high frequency and requires only a "TA" dinucleotide as the essential target in the recipient DNA. Therefore, it can insert nearly random in all regions of the *Bacillus* chromosome (Le Breton *et al.* 2006).

Screening of a mutant library generated by TnYLB-1 transposon will be done to identify the genes involved in rhizosphere competence (swarming ability and biofilm formation) as well as in plant growth-promoting activity. In addition, colonization of *B. amyloliquefaciens* FZB42 and its mutants on the roots of the plant will be monitored using SEM and CLSM to find out whether or not there is different pattern of colonization. The use of transposon mutagenesis will also be applied to discover novel secondary metabolites by screening the transposon library for mutants impaired in synthesis of antibiotics and in nematocidal activity.

2. Materials and Methods

2.1 Chemicals and materials

All chemicals and materials used in the present study are listed in table 1.

Table 1. Chemicals and materials used in the present study

Manufacturer	Product
Amersham	[γ- 32P]ATP, Plus One Tris-Base, Plus One EDTA, Plus One boric acid
Pharmacia	Ready to Go DNA labelled Beads
BD	Difco medium 3
Biorad	Blotting grade blotter non-fat dry milk
Bioron	Taq polymerase
Fermentas	DNA markers, dNTPs, prestained protein ladder, RevertAid M-MuLV reverse transcriptase (200U/μl), restriction endonucleases, RiboLock ribonuclease inhibitor (40U/ μl), T4 DNA ligase, T4 kinase, T4Polynucleotide kinase
Fluka	$CaCl_2$, EDTA
Macherey-Nagel	Nitrocellulose membrane porablot NCL, Nucleo Spin ® Extract II, Nucleo Spin RNA L, Porablot NY plus, Protino® Ni-1000 kit
Merck	Meat extract
MP	Biomedicals Urea pure
Promega	BCIP (50 mg/ml), NBT (50 mg/ml), pGEM-T® Vector systems
Qiagen	QIAEX II gel extraction kit, QIAprep Spin mini prep kit, Qiaquick PCR purification kit
Roche	Anti-DIG AP, Ampicillin, blocking reagent, DIG-dUTP, kanamycin
Roth	Agarose, chloramphenicol, citric acid, $CuSO_4$, DEPC, $FeCl_2$, $FeCl_2$, $Fe_2(SO_2)_3$, formaldehyde, L-glutamic acid, glycerol, HEPES, IPTG, KCl, K_2HPO_4, H_2KPO_4, maleic acid, $MgSO_4$, $MnCl_2$, $MnSO_4$, Na-acetate, Nacitrate, Na_2CO_3, NaCl, NaOH, $(NH_4)_2SO_2$, peptone, SDS, Proteinase K, Rotiphorese Gel 40 (19:1), Rotiphorese Gel 40 (29:1), TEMED, Tris, Triton-X 100, Tween 20, XGal, yeast extract, $ZnCl_2$
Serva	Agar, APS, boric acid, casamino acids, DTT, EGTA,

	Erythromycin, glucose, N-Lauroylsarcosine-sodium, lincomycin/HCl, $MgCl_2$, MOPS, NaN_3, Na_2SO_4, ONPG, L-tryptophan
Sigma	Oligonucleotides, Anti-rabbit IgG AP

2.2 Plasmids, bacterial strains and primers

The plasmids, bacterial strains and primers used in this study are listed in tables 2, 3, 4 respectively.

Table 2. Plasmids used in the present study

Plasmid/reference	Description
pMarA/Le Breton et al. 2006	pUC19 carrying TnYLB-1 transposon, mariner-Himar1 transposase and promoter σ^A, Kan^r Amp^r Erm^r
pMarB/Le Breton et al. 2006	pUC19 carrying TnYLB-1 transposon, mariner-Himar1 transposase and promoter σ^B, Kan^r Amp^r Erm^r
pMarC/Le Breton et al. 2006	pUC19 carrying TnYLB-1 transposon, Kan^r Amp^r Erm^r
pUC18/Fermentas	Cloning vector Amp^r, *lacZ'*
pVBF	pUC18 carrying fragment of *amyE*
pAB1	pVBF carrying fragment of *pabB*
pAB2	pVBF carrying fragment of *yusV*
pAB3	pVBF carrying fragment of *degU*
pAB6	pVBF carrying fragment of *nfrA*
pAB7	pVBF carrying fragment of RBAM_017410
pAB8	pVBF carrying fragment of *abrB*

Table 3. Bacterial strains used in the present study

Strain	Genotype	Reference
B. amyloliquefaciens FZB42	Wild type	FZB Berlin
B. subtilis 168	trpC2	Laboratory stock
E. coli DH5α	supE44 ΔlacU169(Φ80 lacZΔM15) hsdR17 recA1 gyrA96 thi-1 relA1	Laboratory stock
CH5	FZB42 sfp::ermAM yczE::cm	X.-H.Chen, 2009
AB101	FZB42 pabB::TnYLB-1	This Study
AB102	FZB42 yusV::TnYLB-1	This Study
AB103	FZB42 degU::TnYLB-1	This Study
AB106	FZB42 nfrA::TnYLB-1	This Study
AB107	FZB42 RBAM_017410::TnYLB-1	This Study
AB108	FZB42 abrB::TnYLB-1	This Study
AB110	CH5:: TnYLB-1	This Study

Table 4. Primers used in this study

Primer (restriction site)	Sequence (5' to 3' end)	Source or reference
oIPCR1 oIPCR2 oIPCR3	GCTTGTAAATTCTATCATAATTG AGGGAATCATTTGAAGGTTGG GCATTTAATACTAGCGACGCC	Le Breton et al. 2006
yusV-dw-Eco91I yusV-up- SacII	CTCCCTTTGGAATTTGGACAGCCGCTATGAC AGCCCGCGGTCCGTGTATTTCTCAAGCAGG	This work

nfrA-dw- Eco88I nfrA-up- ClaI	AAT<u>CCCGAG</u>ATCGAATCGTTTCATTCCTCG TTA<u>TAGCTA</u>TTCACACCTTCCAGAACATCG	This work
410-up- sacII 410-dw- eco91I	AAC<u>CCGCGG</u>ATTGCATTGAACGGCGGTCT GCA<u>CCATTGG</u>ATCCCTTTGGTATCCCTCAG	This work
degU-dw-ClaI degU-up- Eco88I	AAT<u>ATCGAT</u>TCACCGAAAACCACTTGGAG ATA<u>CCCGAG</u>TAGGATAAGGAGGCGTAGCG	This work
pabB-dw- Eco91I pabB-up- SacII	TTT<u>GGTTACC</u>TGAATAGAGACATACACACGGC AAT<u>CCGCGG</u>ATTCCGTCTGACGATCAGTTC	This work
abrB-dw-SacII abrB-up- Eco91I	TTT<u>CCGCGG</u>AAGAGCATGTGGAGCATTAC GGC<u>CCATTGG</u>AACCTCCCATTCAGAATGTC	This work
amyBack-1 amyBack-2	AGCGAAATTACCTGACGGCAG AGCTCAAGTTCCGTCACACCTG	Ben et al. 2011
amyFront(aatII)-1 amyFront-2	AGTTT<u>GACGTC</u> TCTCCGATTTCGCCGACAACAC TCGATTTGTTTGCAGTTTCAGCG	Ben et al. 2011

2.3 Media and supplements

All media used in this work were prepared and sterilized according to Sambrook *et al.* 1989 and Cutting and Horn 1990. Supplements with different antibiotics and compounds are listed in table 5. For screening biofilm formation, bacteria were grown in MSgg medium (Branda *et al.* 2004). Cultivation of *L. minor* was done in Steinberg medium (Idris *et al.* 2007).

- **LB (Luria-Broth) medium**

1 % w/v   peptone

0,5 % w/v   yeast extract

0,5 % w/v   NaCl

- **MSgg medium**

5 mM	K_2HPO_4 [pH 7]	1 µM	$ZnCl_2$
2 mM	$MgCl_2$	700 µM	$MnCl_2$
50 µM	$FeCl_2$	2 µM	thiamine
0.5%	glycerol	0.5%	glutamate
50 µg/ml	tryptophan	50 µg/ml	phenylalanine
100 mM	morpholinepropane sulfonic acid [pH 7]		

- **Steinberg medium**

3.46 mM	KNO_3	1.25 mM	$Ca(NO_3)_2$
0.66 mM	KH_2PO_4	0.072 mM	K_2HPO_4
0.41 mM	$MgSO_4$	1.94 µM	H_3BO_3
0.63 µM	$ZnSO_4$	0.18 µM	Na_2MoO_4
0.91 µM	$MnCl_2$	2.81 µM	$FeCl_3$
4.03 µM	EDTA		

Table 5. Supplements

Supplement	Final concentration
Agar	1,5 % w/v, 0,75 % w/v (swarming agar plates)
Amplicillin	100 µg/ml

Chloramphenicol	20 µg/ml (for *E. coli*), 5 µg/ml (for *Bacilli*)
Erythromycin	1 µg/ml (for *Bacilli*)
IPTG	1 mM
Kanamycin	20 µg/ml (for *E. coli*), 5 µg/ml (for *Bacilli*)
Lincomycin	25 µg/ml (for *Bacilli*)
XGal	40 µg/ml

2.4 Molecular Biology techniques

2.4.1 Standard molecular biology methods

DNA manipulation, such as digestion with restriction endonucleases and ligation, was performed according to the instructions supplied by the manufacturer. Agarose-gel electrophoresis, fluorescent visualization of DNA with ethidium bromide, spectrophotometric quantitation of DNA as well as preparation of $CaCl_2$-competent *E. coli* cells followed by transformation of plasmid DNA were carried out with standard procedures described by Sambrook *et al.* 1989. Bacterial chromosomal DNA from *Bacilli* was prepared as described by Cutting and Horn 1990b. Polymerase chain reaction (PCR) was done using the GeneAmp PCR system 2700 (Applied Biosciences) according to Dieffenbach and Dveksler 1995, under the appropriate conditions in each case. Ligation of PCR products to pGEM-T vector was carried out following the instructions of the manufacturer (Promega). Plasmid DNA isolation and recovery of DNA from agarose gels were performed with QIAprep Spin mini prep kit and QIAEX II gel extraction kit, respectively.

2.4.2 Transformation in *B. amyloliquefaciens*

Competent cells of *Bacillus amyloliquefaciens* were obtained by modifying the two-step protocol published by Kunst and Rapoport 1995. Cells were grown overnight in LB medium at 28°C (170 rpm). The next day, they were diluted in glucose-casein hydrolysate-potassium phosphate (GCHE) buffer to an OD_{600} of 0,3. The cell culture was then incubated at 37°C under vigorous shaking (200 rpm) until the middle of exponential growth (OD_{600} ~1,4). Dilution with an equal volume of GC medium followed and the cells were further incubated under the same conditions for 1 hour. Further on, the culture was divided in 2 ml Eppendorf tubes and cells were harvested by centrifugation at 6000 rpm for 5 minutes (room temperature). The pellets were resuspended in 200 µl of the supernatant and the desired DNA (1 µg) with 2 ml transformation buffer was added to them. After incubation at 37°C under shaking at 75 rpm for 20 minutes, 1 ml LB medium containing sublethal concentration (0,1 µg/ml) of the appropriate antibiotic was added. The cells were grown under vigorous shaking for 90 minutes and platted on selective agar plates.

Buffers

●GCHE buffer	●GC buffer
1 x PC buffer	1 x PC buffer
0,1 M glucose	0,1 M glucose
0,005% w/v tryptophan	0,005% w/v tryptophan
0,04 M $FeCl_3$ / Na-citrate	0,04 M $FeCl_3$ / Na-citrate
0,25% w/v potassium glutamate	3 mM $MgSO_4$

3 mM MgSO$_4$

0,1% w/v casein hydrolysate

●10 x PC buffer	●Transformation buffer
0,8 M K$_2$HPO$_4$	1 x SMM buffer
0,45 M H$_2$KPO$_4$	1 mM EGTA
0,028 M Na-citrate	0,025 M glucose
	0,02 M MgCl$_2$

2.4.3 Transposon mutagenesis

2.4.3.1 Detection of *mariner* transposition events

The *mariner* based transposon TnYLB-1 plasmid was used to generate a transposon library according to Haldenwang (Le Breton *et al.* 2006). In brief, plasmid pMarA, pMarB and pMarC were transformed into *B. amyloliquefaciens* FZB42 selecting for Kan^r at 30°C. Transformant colonies were screened for plasmid-associated properties, i.e. Kan^r and Erm^r at permissive temperature for plasmid replication (30°C) and Kan^r and Erm^s at the restrictive temperature (48°C). Plasmid DNA was then extracted from the transformants and subjected to restriction endonuclease analysis to verify that these transformants contained the original intact plasmid. Then representative plasmid-containing colonies were incubated overnight in LB medium at 37°C. Samples were then plated on LB agar containing Kan and incubated at 48°C to select for transposants.

2.4.3.2 Mapping of transposon insertion sites

Five micrograms of genomic DNA isolated from the respective transposants was digested with *Taq* I and then circularised in a ligation reaction using 'Rapid Ligation' kit (Fermentas, Germany) at a DNA concentration of 5 ng/μl. Inverse PCR was performed on 100 ng of ligated DNA using oIPCR1 and oIPCR2, which face outward from the transposon sequence. IPCR products were purified using PCR purification kit (Amersham, UK) and sequenced using the primer oIPCR3 (Le Breton *et al.* 2006).

2.4.4 Hybridization analysis of southern blots

Southern blot is a way of permanently immobilizing DNA (that has been separated by agarose gel electrophoresis) to a solid support. It is designed to locate a particular sequence of DNA within a complex mixture, such as an entire genome. Hybridization and detection occurs by "anealling" with a complementary labelled DNA probe.

2.4.4.1 Synthesis of DIG-labelled probe

For each Southern hybridization, an appropriate probe was labelled with Digoxigenin-11-dUTP (DIG-dUTP), according to the Ready-to-Go kit from Roche. The desired DNA region was amplified by PCR and purified, prior to labelling. 100 ng of the PCR fragment were denaturated by heating at 100°C for 10 minutes and then mixed with 5 μl dCTP (10 mM), 2,5 μl DIG-dUTP (1mM) to a final volume of 50 μl. The mixture was incubated at 37°C for 1,5 hours and was stored at -20°C until use.

2.4.4.2 Preparation of samples; transfer and fixation on a membrane

1-2 μg of the chromosomal DNA in question were digested overnight with a suitable restriction endonuclease. Samples were initially separated on a 0,8 % agarose gel in 1 x TAE buffer at 70 Volt. The gel was washed twice for 20 minutes, initially with denaturation buffer and subsequently with neutralization buffer. Transfer on a nylon membrane was performed using the Biorad vacuum blotter (model 785). The DNA was fixed permanently on the membrane by cross-linking using UV radiation.

Buffers

●Denaturation buffer	●Neutralization buffer
1,5 M NaCl	1,5 M NaCl
0,5 M NaOH	1 M Tris-HCl pH=8.0

2.4.4.3. Hybridization and detection

The membrane was initially incubated for 1 hour at 65°C with 40 ml hybridization buffer and was hybridized overnight at 55°C with 5-10 ml hybridization buffer containing 5-25 ng/ml of denaturated DIG-labelled probe. The membrane was washed twice for 15 minutes, first with 2 x SSC/0,1 % SDS at room temperature and then with 0,5 x SSC/0,1 % SDS at 55°C. Detection was achieved by a colorimetric approach. The membrane was first equilibrated with P1-DIG buffer and was then incubated for 30 minutes with P1-DIG buffer containing 3,75 units of the antibody Anti-Digoxigenin-Alkaline-Phosphatase. Unbound antibody was removed after a fifteen minute washing step. Addition of 10 ml Ap buffer containing 2,25 mg nitroblue tetrazolium salt (NBT) and 1,75 mg 5-bromo-4-chloro-3 – indolyl phosphate (BCIP) to the membrane and incubation in the dark allowed visualization of the hybridized DNA with our labelled probe.

Buffers

●Hybridization buffer   ●20 x SSC

5 x SSC

1 % w/v blocking reagent

0,1 % v/ N-lauroylsarcosine-sodium

0,02 % w/v SDS

3 M NaCl

0,3 M Na-citrate

•P1-DIG buffer	•Wash buffer	•Ap buffer
0,1 M maleic acid	0,1 M maleic acid	0,1 M Tris-HCl pH=9.5
0,15 M NaCl	0,15 M NaCl	0,1 M NaCl
1 % w/v blocking reagent	0,3 % v/v Tween-20	0,05 M $MgCl_2$

2.5 Screening for plant growth promotion mutants using the Lemna biotest system

L. minor ST was propagated in Steinberg medium. Four plants with two or three budding-pouches (fronds) were incubated in 200 ml of medium in a 500 ml flask. The flasks were kept at 22°C with continuous light until sufficient numbers of homogenous *Lemna* plant were obtained. The growth medium was changed every week. To prove the biostimulation effect of FZB42 a 48-well microtiter plate was used. Each well was filled with 1.25 ml of Steinberg medium. *Lemna* plants with two fronds were transferred aseptically into the microtiter plates. Culture transposon mutant in appropriate dilutions were added directly. The microtiter plates were kept at 22°C and 24 h light for 10 days, Plants were harvested and growth was determined by dried weight. The result of each trial was repeated four times (Idris *et al.* 2007).

2.6 Assay for plant growth promotion with *Arabidopsis thaliana*

2.6.1 Sterilisation of *Arabidopsis* seeds

Arabidopsis seeds (*Arabidopsis thaliana* var. Columbia) were transferred to an Eppendorf tube and added with 1 ml of 10% sodium hypochlorite. The tube was then shaken for 3 minutes. After pipetting off the sodium hypochlorite solution, 1 ml of sterile distilled water was added to remove residual sodium hypochlorite from the seeds. The tube was inverted 5 times to ensure thorough washing of the seeds. The water was removed with a pipette and repeated the sterile water wash 4 times. The majority of the water was removed, leaving a small volume in the base of the tube to facilitate plating of seeds.

2.6.2 Plant growth conditions

Surface sterilized seeds were pre-germinated on petri dishes containing medium consisting of half-strength Murashige and Skoog 0.6% agar and 3% sucrose and allowed to germinate for 7 days at 22°C. The roots of seven-days-old Arabidopsis seedlings were dipped into the bacterial suspension (1×10^5 CFU/ml) for 5 min and four seedlings were transferred into square petri dish containing half-strength Murashige and Skoog medium with 1% agar. The square petri dishes were placed in a growth chamber at 22°C with 14-h photoperiod. Fresh weight of the plants was measured at 21 days after transplanting.

2.7. Screening for biofilm, swarming and antibiotic mutants

2.7.1 Screening of biofilm mutants

Transposon mutants of *B. amyloliquefaciens* were inoculated in 140 µl of LB medium containing kanamycin within a 96-well microtiter plate. The microtiter plates were shaken at low speed (160 rpm) at 37°C for 16 h. Then, 5 µl of every culture were transferred into 1 ml MSgg medium containing kanamycin within a 48-well microtiter plate. The microtiter plates were incubated without shaking at 30°C for 60 h and development of biofilms was analyzed by visual inspection (Branda et al. 2004).

2.7.2 Screening of swarming mutants

Transposon mutants of *B. amyloliquefaciens* were inoculated, 25 at a time, into LB plus kanamycin solidified with 0.9% agar and incubated at 30°C overnight. Putative swarming mutants were indentified as small colonies and picked into individual 30 mm diameter plates containing 5 ml of swarm agar (LB solidified with 0.7% agar) supplemented with kanamycin and incubated at 30°C overnight. The mutants that remained unable to completely colonize the mini plates were then verified under the standard conditions for swarming motility by inoculating on LB swarm agar containing kanamycin and incubated 24 h at 37°C (Kearns et al. 2004).

2.7.3 Screening of antibiotics mutants

Transposon mutants of *B. amyloliquefaciens* were inoculated in 2 ml LB medium and incubated until OD 1 was reached. At the same time, *B. subtilis* HB0042 was incubated in 10 ml LB medium until OD 0.6 was reached. The culture of *B. subtilis* HB0042 was then poured in LB agar handwarm (1:40 dilution). The mixture was poured in plate and let to dry. Two ul of transposon mutant was inoculated on the plate and incubate overnight at 37°C.

3. Results

3.1 Transformation *B. amyloliquefaciens* FZB42 with the transposon plasmid TnYLB-1

B. amyloliquefaciens FZB42 is known as a plant growth promoting rhizobacterium because it offers not only protection towards the competitive plant-pathogenic microflora within rhizosphere by secretion of antifungal and antibacterial lipopeptides and polyketides (Koumoutsi *et al.* 2004; and Chen *et al.* 2006) but also by production of plant hormones such as IAA (Idris *et al.* 2007). However, the molecular mechanisms behind this ability are not fully understood. In order to find out the beneficial action of this strain at molecular level, transposon mutagenesis was performed.

Transposon-based mutagenesis is a powerful technique for generating mutant libraries, and its use has led to the identification of gene functions in various bacterial systems. In bacteria, transposons are widely employed as random insertion mutagens both at a genome level or and in the analysis of the organization of individual genes (Hayes, 2003; and Picardeau, 2010). Mariner-family transposable elements are a diverse and taxonomically widespread group of transposons occurring throughout the animal kingdom. Among hundreds of different mariners, only two are known to be active, these are *Mos1* and *Himar1*. Both require no host-specific factors for transposition and so have been advanced as generalized genetic tools (Lampe *et al.* 1999). Himar1 has been used as a prokaryotic genetic tool such as in *Burcella melitensis* (Wu *et al.* 2006), *Leptospora interrogans* (Bourhy *et al.* 2005), *Leptospira biflexa* (Louvel *et al.* 2005), *Rickettsia prowazekii* (Liu *et al.* 2007), *Borrelia burgdorferi* (Morozova *et al.* 2005), and *Bacillus subtilis* (Lebron *et al.* 2006).

In this research three different plasmids containing a mariner-based *Himar1* tranposon namely pMarA, pMarB and control plasmid pMarC were used in transposon mutagenesis in *B. amyloliquefaciens* FZB42. Plasmid pMarA and pMarB differ in the promoters that drive the expression of the *Himar1* transposase gene. pMarA has *Himar1* under the transcriptional control of housekeeping σ factor $σ^A$ of *B. subtilis*, while pMar B uses general stress response σ factor $σ^B$ for transposase expression. pMarC has no transposase gene as well as its promoter and is used as a control (Le Breton *et al.*2006).

Transformation of these plasmids was done by modification of the method of Kunst, F and Rapoport, G. (1995). The same amount (1 μg) of plasmid DNA from pMarA, pMarB and pMarC was transformed into FZB42 (see material and methods). pMarA and pMarC have been successfully transformed into FZB42, however, pMarB failed. Because pMarA and pMarB contained the same *Himar1* mariner transposase gene only differing in their respective promoters, we continued to use the pMarA as a source of transposon mutagenesis.

Transformants that contained plasmid pMarA had to be verified that they contained the original intact plasmid before being used for transposon mutagenesis. This was done by screening the transformants for the plasmid-associated properties, i.e. Kan^r and Ery^r at the permissive temperature for plasmid replication (30°C) and Kan^r and Ery^s at the restrictive temperature (48°C). Then the plasmid was extracted from the transformants and transformed into *E. coli* DH5α. Next, plasmid DNA was extracted from *E. coli* DH5α and subjected to restriction endonuclease analysis with *EcoRI*. The restriction was then analysed through agarose gel electrophoresis to verify that the transformants contained the correct plasmid. Fig. 2 shows the restriction analysis of plasmid extracted from transformants *E. coli* DH5α and plasmid pMarA as a positive control.

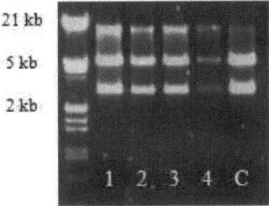

Fig. 2. Restriction analysis of plasmid DNA cut with *EcoRI*. Lane 1-4 from transformed *E. coli*, lane C from plasmid pMarA.

3.2 *Himar1* transposon mutagenesis of *B. amyloliquefaciens* FZB42

After verifying that the plasmids pMarA and pMArC were correctly inserted in *B. amyloliquefaciens* FZB42, the transposon mutagenesis was done by growing the isolated clones overnight in liquid LB medium at 37°C. Then each culture was plated either on LB, LB plus 5 μg/ml Kan or LB plus 1 μg/ml Ery and incubated at the nonpermissive temperature for plasmid replication (48°C). Representative data that are the average data of two separate experiments are presented in Table 6. Kan^r clones represented in this transposition events appeared at frequency ~ 10^{-2} which is significantly higher than that reported for transposons Tn917 and Tn10 (10^{-6} and 10^{-4}, respectively), which are commonly used in *B. subtilis*. There are no antibiotic-resistant clones detected when *B. amyloliquefaciens* FZB42 carrying pMarC lacking of transposase coding sequence was plated in LB plus Kan and LB plus Ery at 48°C. Hence, the Emr^r clones detected were likely a consequence of transposition event from plasmid multimers in which most plasmid sequence were inserted into the *B. amyloliquefaciens* FZB42 chromosome (Le Breton *et al.* 2006).

Table 6. Average of transposon frequency

Delivery Plasmid	Viable cell count (CFU/ml)			Transposition frequency	Erm^R/Kan^R
	LB 48°C	LB Kan^R 48°C	LB Erm^R 48°C		
pMarA	3.4 x 10^8	2.6 x 10^7	3.6 x 10^5	7.6 x 10^{-2}	0.22%
pMarC	2.5 x 10^8	0	0	-	-

Southern blot analysis was done to verify integration of the transposon and to test whether the insertions are likely to be random. In this analysis chromosomal DNA from *B. amyloliquefaciens* FZB42 and clones were isolated and digested with EcoRI. Digoxigenin-labeled DNA specific for the transposon was created by cutting TnYLB-1 region with *PstI*. Hybridization of this probe to EcoRI-digested DNA from clones gave the patterns illustrated in Fig. 3 A. In addition, PCR for the presence of the kanamycin resistance gene was also performed using primers flanking the kanamycin coding sequence in all clones including their complementation and retransformation (Fig. 3 C).

A.

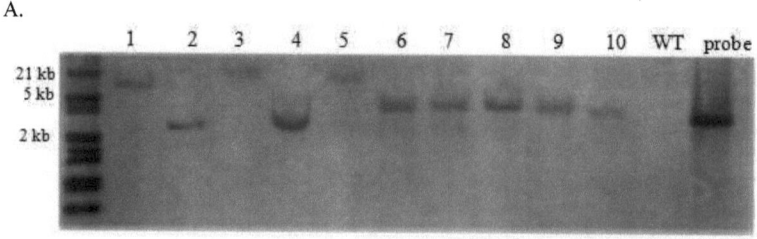

B.

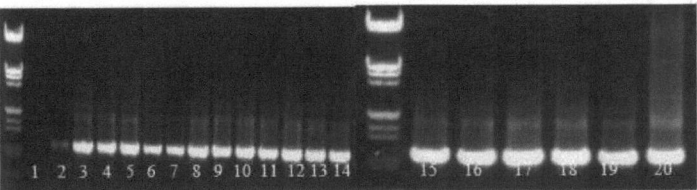

Figure 3. TnYLB-1 transposition in *B. amyloliqufaciens* FZB42. A. Southern hybridization analysis of randomly chosen *B. amyloliquefaciens* FZB42 TnYLB-1 insertion mutants. Chromosomal DNA from *B. amyloliquefaciens* FZB42 (WT) and transposants (lanes 1-10) were digested with EcoRI and analyzed by Southern blotting using a hybridization probe specific for TnYLB-1. DNA fragment sizes (kbp) areindicated to the left and are based on DNA markers. B. PCR products of kanamycin gene, wild type FZB42 (lane 1) and the mutants (lane 2-20).

3.3 Mapping of transposon insertion mutants

To verify that transposition with TnYLB-1 is an efficient tool to get insertion mutations, 787 temperature-resistant Kan^r clones were spotted onto glucose minimum medium to screen for auxotrophic mutations. Of these, seven (~1%) clones spotted failed to grow on the minimal medium, indicating that the transposon insertion was an effective way to create mutation. To identify the *B. amyloliquefaciens* FZB42 genes disrupted by insertion and to further characterize the insertion sites, chromosomal DNA was extracted from the auxotroph phenotype. Two DNA samples isolated from auxotroph phenotype were used in an inverse PCR protocol using primers oIPCR1 and oIPCR2 that allows amplifying the flanking region of the transposon containing the inverse terminal repeat. The amplified DNAs were then sequenced using primer oIPCR3 and the sequences were characterised by BLAST analysis (Le Breton *et al.* 2006). Each of the two auxotroph mutants that were examined yielded in an insertion at an unique location on the *B. amyloliquefaciens* FZB42 chromosome. The two putative insertions were found in *pabB* encoding para-aminobenzoate synthase (subunit A) and *hisJ* encoding histidinol phosphate phosphatase. Additions of histidin and para aminobenzoic acid in the minimal medium restore the growth of *hisJ* and *pabB* auxotroph, confirming the need of those compounds for growth of both auxotrophic mutants.

B. amyloliquefaciens FZB42 which contains plasmid transposon TnYLB-1 was then used to create mutant library of the different phenotype, i.e. mutant in biofilm production, swarming, plant growth promotion, nematocidal production and antibiotic. Three biofilm mutants (*pabB, yusV* and *degU*), one swarming mutant (*degU*), three plant growth promotion mutants (*nfrA, abrB* and *RBAM_017410*), five nematocidal mutants (*prkA, yhdY, ywmL, mlnD* and *RBAM_007470*) and five antibiotic mutants (*degU, degS, nrsB, yaaT* and *RBAM_029230*) were found in this screening (Romy Scholz and Quisheng He, unpublished data). These insertion sites are mapped onto a circular representation of the *B. amyloliquefaciens* FZB42 genome in Fig. 4, revealing a genome-wide distribution of insertion sites that are found in open reading frames. Collectively, the data shown in Fig. 3 and 4 indicate that transposition of the TnYLB-1 element in *B. amyloliquefaciens* FZB42 produces predominantly mutants with single insertions that are distributed around the chromosome in an apparently random fashion.

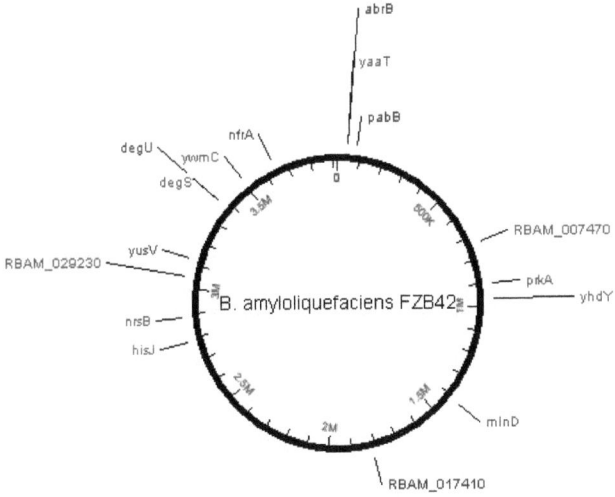

Figure 4. Random distribution of TnYLB-1 insertions in the *B. amyloliquefaciens* FZB42 chromosome. Brown color represents biofilm mutants, blue color represents PGPR mutants, green color represents nematocidal mutants, red color represents antibiotics mutants and grey color represents auxotrophic mutants.

3.4 Discovery of genes involved in swarming motility and biofilm formation

Bacteria motility mechanisms, including swimming and swarming, can have a profound impact on the colonization of surfaces, the first step in the formation of adherent microbial assemblies called biofilms. Both features are known to play important role on how prokaryotes interact with and adapt to surface environments (Mireles II *et al.* 2001; and Meritt *et al.* 2007). To discover the genes involved in swarming motility and biofilm formation, I carried out insertional mutagenesis in *B. amyloliquefaciens* FZB42 using the transposon TnYLB-1. Approximately 6000 colonies from the transposon library were screened for being impaired in swarming motility on swarming agar plates (LB solidified with 0.7% agar). Swarming motility of *B. subtilis* has previously been evaluated with this approach (or slight variations thereof) by several groups (Senesi *et al.* 2002; Kearns *et al.* 2003; and Calvio *et al.* 2005). One mutant in swarming motility was found in this screening which was impaired in its ability to cover the whole area of swarming agar plate (Fig. 8). Later, on biofilm screening this mutant also showed a defect in biofilm formation.

I screened the TnYLB-1 insertion library using MSgg media to find biofilm mutants. Defect in biofilm formation of *B. subtilis* has previously been screened with this approach (Brenda *et al.* 2004). Using this assay, 6000 mutants were screened to identify mutants exhibiting a reduced ability to form biofilms. Three unique mutants that were impaired in biofilm production in MSgg media were isolated. To identify the *B. amyloliquefaciens* FZB42 genes disrupted by the insertions, chromosomal DNA from these three isolates was extracted and used in an inverse PCR protocol that amplified the chromosomal DNA abutting the transposons's inverse terminal repeats (ITRs). The amplified DNA was then sequenced using primer oIPCR3. BLAST analysis of the DNA sequences revealed that these mutants carried TnYLB-1 insertions at distinct genes. The insertion of TnYLB-1 transposon in two mutants that showed a reduced ability to form biofilm was in the *yusV* and *pabB* genes, whereas the mutant that exhibited defect in swarming ability and biofilm formation was in the *degU* gene. The *yusV* gene has unknown function, but it is similar to iron (III) dicitrate transport permease and *pabB* gene encodes for para-aminobenzoate synthase (subunit A). These three mutants showed different phenotypes in respect to biofilms formation. The mutant of *B. amyloliquefaciens* FZB42 harboring *degU*::TnYLB-1 formed thin pellicle when grown in standing liquid MSgg medium (Fig. 9 B), the mutant harboring *yusV*::TnYLB-1 formed pellicles with flat surface (Fig. 13 B), whereas the mutant harboring *pabB*::TnYLB-1 formed flat and thin pellicles with granules on the top of the surface (Fig. 17 B).

3.4.1 *B. amyloliquefaciens* FZB42 *degU*::TnYLB-1

Nucleotide sequence analysis revealed that the transposon insertion in the mutant which was defect in biofilm formation and swarming motility took place within *degU* gene. The *degU* has a key role in regulating several post-exponential phase processes in *B. subtilis*, including the activation and inhibition of genetic competence, the inhibition of flagellar based motility, the activation of degradative enzyme production and the activation of poly-γ-glutamic acid production (Verhamme *et al.* 2007). Therefore, the *degU* mutant is a regulatory mutant, because it involves in regulating many bacterial cell activities. The genetic organizations of the *degU* regions carrying transposon insertion are shown in Fig. 5.

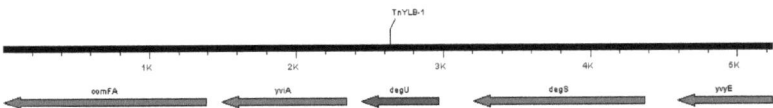

Figure. 5. Genomic organization of *degU* region carrying the TnYLB-1 insertion and its flanking regions.

3.4.1.1 Complementation of *degU* gene

In order to confirm that insertion of TnYLB-1 *mariner*-based tranposon in *degU* gene was responsible for impairing swarming motility in *B. amyloliquefaciens* FZB42, complementation of the *degU* mutant by inserting the intact *degU* was done. To complement the *degU* gene, the coding region plus 178 bp of upstream sequence and 183 bp of downstream sequence were amplified using primers degU-dw-ClaI and degU-up-Eco88I containing Eco91I and SacII site respectively (Fig. 6). The fragment (889 bp) was cloned into linearized *ClaI/Eco88I* pUC18 plasmid which had an *Amy* cassette (pVBF). The ligated DNA was then transformed into $CaCl_2$-competent DH5α cells. To confirm the clones containing the *degU* gene fragment, PCR using degU-dw-ClaI and degU-up-Eco88I primers was performed. The correct plasmid pVBF-ΔdegU cassette was transformed into *degU* mutant. The transformants were then grown in LB plus Kan (5 ug/ml), Ery (1 ug/ml) and 1% amylum. The transformants without clear zone when the medium was added with iodine solution were the the correct transformants.

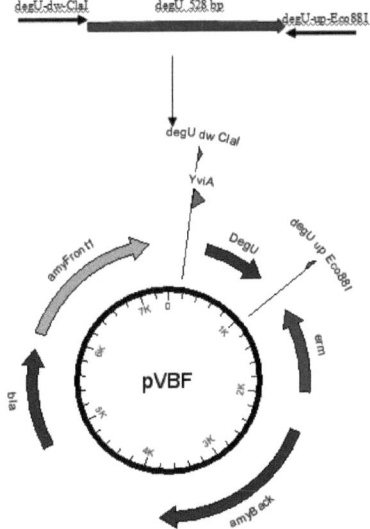

Figure 6. Strategy for construction of pUC18-ΔdegU cassette.

To validate the correct insertion of the complementation and retransformation transformants, chromosomal DNA was amplified by PCR using primers degU-dw-ClaI and degU-up-Eco88I and compared with the *B. amyloliquefaciens* FZB42 and *degU* mutant (Fig. 7). In the complementation transformant, PCR fragment of the *degU* gene showed a similar fragment length as a wild type suggesting that the complementation had occurred, whereas in in retransformation transformant, the length of the *degU* gene was longer than the wild type indicating that *degU* has been inserted with TnYLB-1 segment as in *degU* mutant.

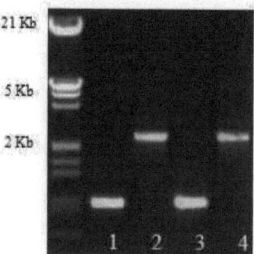

Figure 7. PCR product of *degU* gene. Wild type FZB42 (lane1), *degU* mutant (lane 2), complementation of *degU* (lane 3) and retransformation of *degU* (lane 4).

Insertion of intact *degU* gene into the *degU* mutant restored the swarming motility of the mutant (Fig. 8 C). An additional approach to demonstrate that *degU* plays a role in swarming motility was done by retransformation. In retransformation, chromosomal DNA of *degU* mutant that has been disrupted by insertion of TnYLB-1 was transformed into the wild type of *B. amyloliquefaciens* FZB42. The phenotype of transformants from retransformation showed an impaired swarming motility as in *degU* mutant confirming that the *degU* gene had a role in swarming motility (Fig. 8).

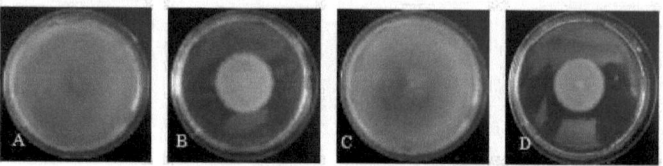

Figure 8. Phenotype of swarming motility in *degU* mutant. *B. amyloliquefaciens* FZB42 (A), *degU* mutant (B), complementation *degU* (C) and retransformation of mutant *degU* (D).

The *degU* gene is also responsible for biofilm formation. The *degU* mutant was impaired in its ability to form robust pellicles as in the wild type when grown in MSgg medium. This mutant formed thin pellicles with little granules in the center (Fig. 9 B). Complementation *degU* mutant with intact *degU* from *B. amyloliquefaciens* FZB42 repaired the mutant ability to produce biofilm in standing MSgg medium (Fig. 9 C). On the contrary, retransformation of *degU*::TnYLB-1 fragment to the wild type of *B. amyloliquefaciens* FZB42 resulted in transformants impaired in biofilm production with phenotype relatively similar to the *degU* mutant (Fig. 9 D). In *Bacillus subtilis* JH642 *degU* was shown to enhance biofilm formation through activating poly-γ-glutamic acid production (Stanley and Lazazzera, 2005). However the role of poly-γ-glutamic acid in *B. subtilis* wild strain 3610 and B-1 in the formation of sessile communities had an inconsistent role (Brenda *et al.* 2006), thus raising the question whether *degU* had a variable role in controlling biofilm formation.

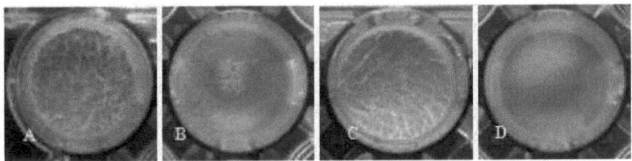

Figure 9. Phenotype of biofilm formation in *degU* mutant. *B. amyloliquefaciens* FZB42 (A), *degU* mutant (B), complementation of *degU* (C), and retransformation of *degU* (D).

3.4.2 *B. amyloliquefaciens* FZB42 *yusV*::TnYLB-1

The insertion of TnYLB-1 tranposon found in the second biofilm mutant was in the *yusV* gene. Its function is unknown but it is similar with iron (III) dicitrate transport permease. The *yusV* gene lies in the *yus* operon which still has unknown function (*yusU*, *yusW*, *yusX* and *yusZ*). Genetic map of insertion TNYLB-1 in *yusV* gene and its neighboring region is shown in Fig. 10.

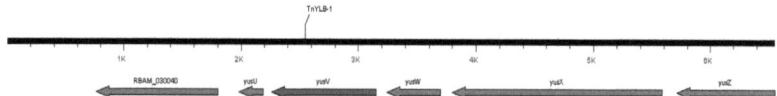

Figure 10. Genomic organization of *yusV* region carrying the TnYLB-1 insertion and its flanking regions.

3.4.2.1 Complementation of *yusV* gene

The intact *yusV* gene plus 77 bp of upstream sequence and 170 bp of downstream sequence was amplified using primers yusV-up-SacII and yusV-dw-Eco91I containing Eco91I and SacII site respectively. The fragment (1072 bp) was then cloned into pUC18 plasmid which had an *Amy* cassette (pVBF) and linearized with *SacII* and *Eco91I* restriction enzymes (Fig. 11). The ligated plasmid DNA was transformed into CaCl$_2$-competent *DH5α* cells. To select the clones with *yusV* insertion, the transformants were grown in the medium containing Amp^r and confirmed with PCR using plasmid DNA as a template with yusV-up-SacII and yusV-dw-Eco91I primers. The plasmid with the correct insertion of *yusV* gene was then transformed into the *yusV* mutant. The transformants were selected for Kan^r and Ery^r with no clear zone in media when it was added with iodine solution. Retransformation of the *yusV*::TnYLB-1 fragment into the wild type was also done in order to confirm that the *yusV* gene was responsible for the defect in biofilm formation in *yusV* mutant.

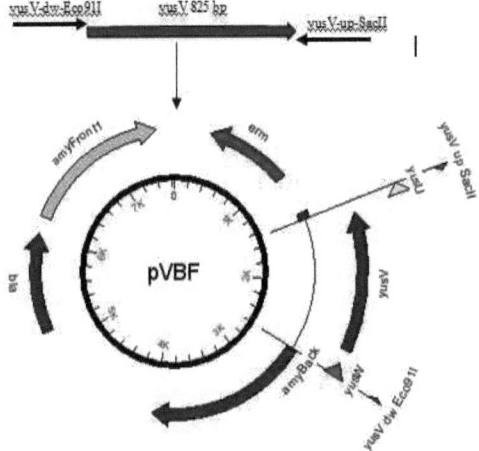

Figure 11. Strategy for construction of pUC18-Δ*yusV* cassette.

To validate that the transformants contained the right insertion, the chromosomal DNA of transformants was amplified using the appropriate primers and compared to the wild type and the mutant. PCR that amplified *yusV* gene in *yusV* mutant showed a longer PCR fragment to the wild type indicating the insertion of transposon TnYLB-1. Two bands of PCR product of *yusV* gene in the complementation confirming that the intact *yusV* gene and the *yusV*::TnYLB-1 fragment were in the chromosome of *yusV* mutant. Retransformation of *yusV*::TnYLB-1 fragment to *B. amyloliquefaciens* FZB42 resulted in a longer PCR fragment in contrast to the wild type (Fig. 12).

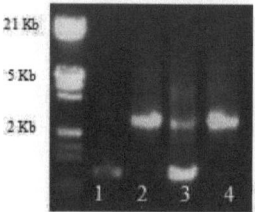

Figure 12. PCR product of *yusV* gene, wild type FZB42 (lane1), *yusV* mutant (lane 2), complementation of *yusV* (lane 3) and retransformation of *yusV* (lane 4).

The finding that the *yusV* gene is involved in biofilm formation has never been reported in any publication (Branda *et al.* 2006; Kobayashi 2007; and Chai *et al.* 2008), hence adds a new insight in biofilm formation. Insertion of TnYLB-1 in *yusV* gene impaired the ability of the mutant to produce biofilm. The *yusV* mutant produced very thin pellicles when grown in MSgg medium (Fig. 13 B), whereas the wild type formed thick pellicles (Fig. 13 A). When complemented with intact *yusV* gene, the insertion restored the ability of the *yusV* mutant to form biofilm (Fig. 13 C). Retransformation of *yusV*::TnYLB-1 fragment to the wild type generated the inability of the wild type phenotype to form a thick pellicles (Fig. 13 D).

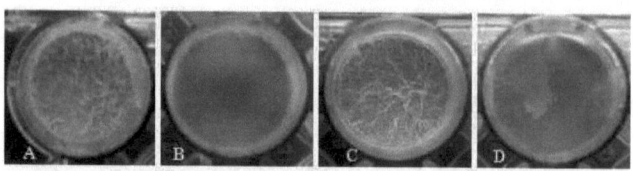

Figure 13. Phenotype of biofilm formation in *yusV* mutant , *B. amyloliquefaciens* FZB42 (A), *yusV* mutant (B), complementation of *yusV* (C), and retransformation of *yusV* (D).

3.4.3 *B. amyloliquefaciens* FZB42 *pabB*::TnYLB-1

The *pabB* gene is one of the folate biosynthesis operon genes, that encodes for para-aminobenzoate synthase. Insertion of TnYLB-1 tranposon in this region generated auxotroph and biofilm mutant. The genomic organisation of the chromosomal region carrying the TnYLB-1 insertion in the *pabB* gene and its flanking region is depicted in Fig. 14.

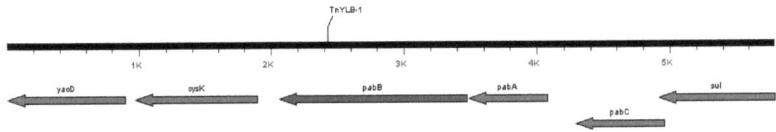

Figure 14. Genomic organization of *pabB* region carrying the TnYLB-1 insertion and its flanking regions.

3.4.3.1 Complementation of *pabB* gene

To complement the *pabB* gene that has been disrupted by the insertion of TnYLB-1 transposon, the plasmid harboring the intact *pabB* gene from *B. amyloliquefaciens* FZB42 was constructed. The construct contained the gene *pabB*/pabB-dw-Eco91I/pabB-up-SacII fragment (1695 bp), which was inserted into linearized *Eco91I/SacII* pUC18 containing *Amy* cassette (pVBF) (Fig. 15). The plasmid bearing the wild type allele was transformed into $CaCl_2$-competent *E. coli DH5α* and selected for Amp^r, which was associated with plasmid transformation. To verify the ligation of the wild type gene in the pVBF plasmid, the wild type gene was amplified by PCR using plasmid DNA extracted from *E. coli DH5α* as the template with pabB-dw-Eco91I and pabB-up-SacII primers.

The plasmid containing the correct insertion of the wild type gene was transformed into *pabB* mutant and selected for Ery^r.

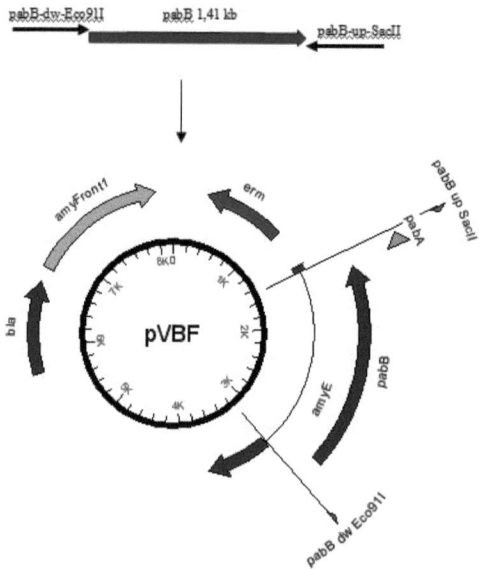

Figure 15. Strategy for construction of pUC18-Δ*pabB* cassette.

The length of PCR fragment which amplified the *pabB* gene in the wild type was different from the *pabB* mutant, complementation and retransformation. The *pabB* mutant, its complementation and retransformation showed the same band of PCR product confirming that the insertion of TnYLB-1 in the *pabB* gene had occured. However, the complementation showed also an extra PCR fragment which has the same size as the wild type indicating that the intact *pabB* gene has been successfully inserted to the mutant (Fig. 16).

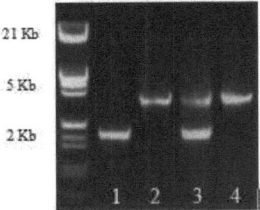

Figure 16. PCR product of *pabB* gene. wild type FZB42 (lane1), *pabB* mutant (lane 2), complementation of *pabB* (lane 3) and retransformation of *pabB* (lane 4).

Compared to the wild type, biofilms in *pabB* mutant showed distinct features such as flat and thin pellicles with little granules on the top of the covered surface of MSgg media (Fig. 17 B). Complementation of the *pabB* gene by insertion of the intact *pabB* gene into the *pabB* mutant restored its ability to produce biofilm. The feature of biofilm after complementation showed thick pellicles as the wild type (Fig. 17 C). However, impairment in biofilm formation in this mutant was not directly related to the gene that was involved in biofilm formation but due to the growth defect in the bacteria as *pabB* mutant was an auxotroph. This was confirmed by the addition of 0.1 mM para-aminobenzoic acid (PABA) in MSgg medium which restored the ability of the *pabB* mutant to form biofilm (Fig. 17 E). Retransformation of *pabB* mutant to *B. amyloliquefaciens* FZB42 impaired its ability to produce biofilm (Fig. 17 D), indicating that the *pabB* fragment has transformed into the wild type.

 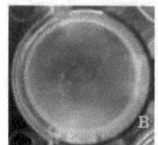 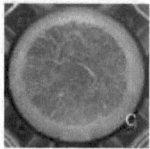 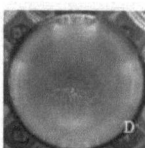

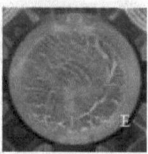

Figure 17. Phenotype of biofilm formation in *pabB* mutant. *B. amyloliquefaciens* FZB42 (A), *pabB* mutant (B), complementation of *pabB* (C), retransformation of *pabB* (D) and addition of 0.1mM PABA (E).

3.5 Discovery of genes involved in plant growth-promoting activity

To investigate the genes that are involved in plant growth-promoting in *B. amyloliquefaciens* FZB42, non-growth-promoting mutants generated by mariner based transposon TnYLB-1 were screened by the Lemna biotest system (Idris *et al.* 2007). In short, 1.25 µl of fresh growing mutant cells at OD600 1 were added to the plant in microtiter plate wells containing 1.25 ml Steinberg growth medium. Each treatment was repeated six times. The plates were kept at 22°C with continuous 24 hours light. After ten days plants were harvested and growth was measured in terms of dry weight. Of 3000 transformants screened, three transformants exhibited impairment in plant growth promoting activity. To identify the genes in these mutants, inverse PCR products were sequenced and searched using BLAST analysis. Blast results showed that the insertion of the mutants were in the *nfrA*, *abrB* and *RBAM_017410* gene.

3.5.1 *B. amyloliquefaciens* FZB42 *nfrA*::TnYLB-1

Since the *nfrA* mutant showed reduction of plant growth-promoting activity of wild type FZB42, I examined whether the complementation of the gene could restore the promoting ability of the mutants. The *nfrA* gene encodes for FMN-containing NADPH-linked nitro/flavin reductase, which is regarded as an essential gene. The region of the *nfrA* gene inserted in TnYLB-1 transposon and its neighboring area is depicted in Fig. 18.

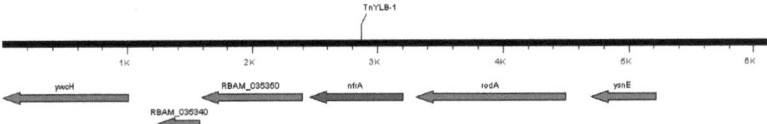

Figure 18. Genomic organization of *nfrA* region carrying the TnYLB-1 insertion and its flanking regions.

3.5.1.1 Complementation of *nfrA* gene

Complementation of the *nrfA* mutant was done by amplifying the *nfrA* coding region plus 241 bp upstream sequences and 278 bp of downstream sequences using nfrA-dw-Eco88I and nfrA-up-ClaI primers, which contained Eco88I and ClaI site, respectively. The fragment of *nfrA*/nfrA-dw-Eco88I/nfrA-up-ClaI (1269 bp) was cloned into linearized *ClaI/Eco88I* pUC18 plasmid which contained an *Amy* cassette (pVBF) (Fig. 19).

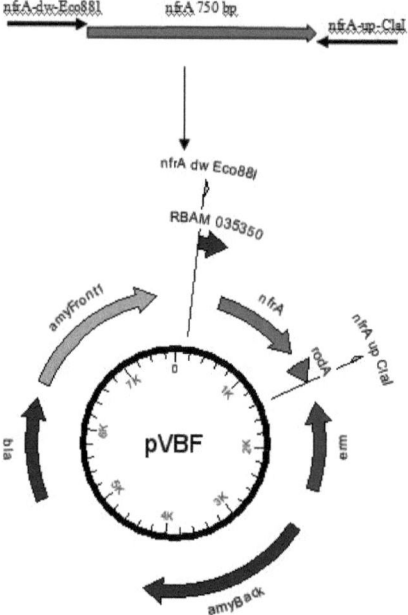

Figure 19. Strategy for construction of pUC18-ΔnfrA cassette.

The pVBF containing fragment of *nfrA*/nfrA-dw-Eco88I/nfrA-up-ClaI was then transformed into CaCl$_2$-competent *DH5α* cells. The clones were selected for Amp^r and confirmed with PCR using the appropriate primers. The plasmid with the appropriate fragment was then transformed to the *nfrA* mutant. Retransformation of the *nfrA*::TnYLB-1 fragment into wild type was done to confirm that the transposon was inserted in this gene. The correct transformants were confirmed with PCR and analysed by gel electrophoresis. Fig. 20 shows the insertion of intact *nfrA* gene into the *nfrA* mutant and insertion of the *nfrA*::TnYLB-1 fragment into the wild type confirming that the complementation and retransformation were successful.

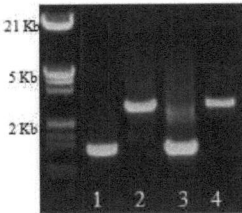

Fig. 20. PCR product of *nfrA* gene. Wild type FZB42 (lane1), *nfrA* mutant (lane 2), complementation of *nfrA* (lane 3) and retransformation of *nfrA* (lane 4).

3.5.1.2 Effect of *nfrA* mutation on growth of *L. minor* and *A. thaliana*

In the *Lemna* biotest system, the *nfrA* mutant showed significant reduction ($p \leq 0.05$) in plant dry weight compared to the wild type. Complementation of *nfrA* gene in mutant restored the plant growth-promoting ability as showed by a significant increase ($p \leq 0.05$) compared to mutant. Retransformation of mutant which caused insertion of *nfrA*::TnYLB-1 into wild type impaired the growth stimulation as indicated by significant reduction ($p \leq 0.05$) compared to wild type (Fig. 21 and 22).

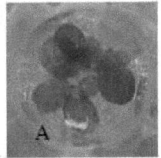
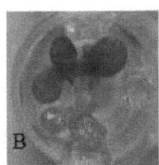
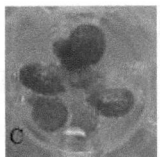
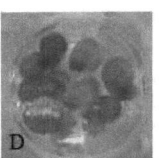

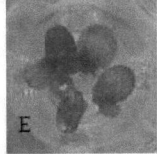

Figure 21. Influence of *nfrA* mutation on plant growth promoting ability of *B. amyloliquefaciens* FZB42 on *L. minor*. The figure shows the representative plant response to the bacterial treatment at 10 days. A. *B. amyloliquefaciens* FZB42, B. Control, C. *nfrA* mutant, D. Complementation, E. Retransformation.

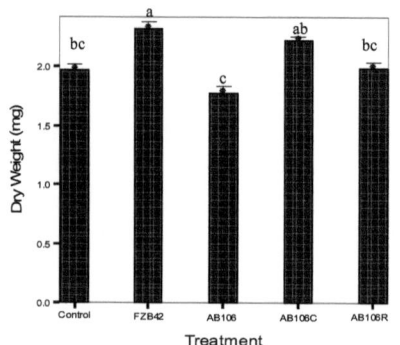

Figure 22. Growth stimulating effects of *nfrA* mutation on *L. minor*. FZB42 : *B. amyloliquefaciens* FZB42, AB106 : *nfrA* mutant, AB106C : Complementation, AB106R : Retransformation. Values shown represent the mean of dry weight ± standard error. Different letters indicates means that differ significantly ($P \leq 0.05$).

In order to examine whether or not the plant growth promotion mutants impaired their ability to promote the growth of other plant than *L. minor*, I tested toward the growth of *Arabidopsis thaliana*. A gnotobiotic system was applied to study the interactions between *B. amyloliquefaciens* FZB42 and *A. thaliana* to exclude other uncontrolled experimental factors that would interfere with this interaction as in natural environment. In this system, the surface sterilized *A. thaliana* seeds were germinated for seven days and the roots were immersed in bacterial culture (1 x 10^{-5} CFU/ml) for five minutes. The plants were then grown in a growth chamber for three weeks. In this assay, the mutant resulted in significant ($p \leq 0.05$) reduction of plant fresh weight compared to wild type. No significant differences were found among control, mutant and retransformation. The wild type and complementation exhibited significant increase ($p \leq 0.05$) compared to control and mutant. These data suggest that complementation restored plant growth promoting activity of the wild type (Fig. 23 and 24).

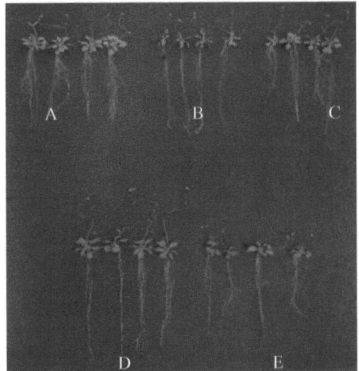

Figure 23. Influence of *nfrA* mutation on plant growth promoting ability of *B. amyloliquefaciens* FZB42 on *A. thaliana*. The figure shows the representative of plant growth to the bacterial treatment at 21 days. A. *B. amyloliquefaciens* FZB42, B. Control, C. *nfrA* mutant, D. Complementation of *nfrA*, E. Retransformation of *nfrA*.

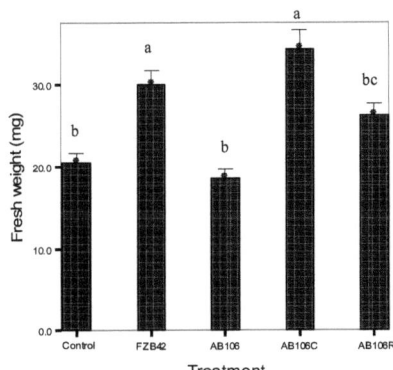

Figure 24. Growth stimulating effects of *nfrA* mutation on *A. thaliana*. FZB42 : *B. amyloliquefaciens* FZB42, AB106 : *nfrA* mutant, AB106C : Complementation, AB106R : Retransformation. Values shown represent the mean of fresh weight ± standard error. Different letters indicate means that differ significantly ($P \leq 0.05$).

3.5.2 *B. amyloliquefaciens* FZB42 *abrB*::TnYLB-1

The *abrB* gene plays a role in regulation of transition state genes which are involved in expression of numerous genes functions such as the formation of biofilm (Hamon *et al.* 2004; and Branda *et al.* 2001), production of extracellular degradative enzymes (Makarewicz *et al.* 2008), initiation of sporulation (Perego and Hoch, 1991) and production of antibiotics (Strauch *et al.* 2008). A mutant with insertion of TnYLB-1 transposon in the *abrB* gene showed reduction in plant growth promoting activity of *B. amyloliqufaciens* FZB42 as detected in screening with *L. minor*. Insertion of TnYLB-1 in the *abrB* coding region and its flanking region is shown in Fig. 25.

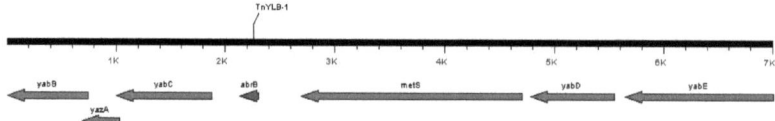

Figure 25. Genomic organization of *abrB* region carrying the TnYLB-1 insertion and its flanking regions.

3.5.2.1 Complementation of *abrB* mutant

To complement the *abrB* mutant, the *abrB* gene plus a 373 bp upstream region and a 396 bp downstream region was amplified using abrB-dw-SacII and abrB-up-Eco91I primers which contained *SacII* and *Eco91I* restriction site, respectively. The fragment (1045 bp) was then cloned into linearized *SacII/Eco91I* pUC18 plasmid which had an *Amy* cassette (pVBF) (Fig. 26). The ligated DNA was transformed into $CaCl_2$-competent *DH5α* cells and selected for Amp^r. The clones containing the *abrB* insertion were confirmed with PCR using abrB-dw-SacII and abrB-up-Eco911 primers. The plasmid was then transformed into *abrB* mutant and selected for Kan^r and Ery^r.

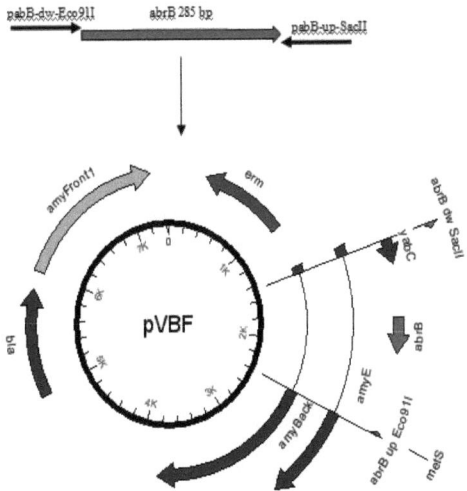

Fig. 26. Strategy for construction of pUC18-ΔabrB cassette.

Amplification of the *abrB* gene in the wild type resulted in a fragment which had a length around 900 bp, whereas in the mutant it was around 2000 bp. Complementation of the *abrB* mutant that has been disrupted by tranposon insertion was done by inserting an intact *abrB* gene from the wild type. The PCR product of *abrB* gene amplification from complementation showed two fragments which indicated that the *abrB* intact gene has been successfully inserted back to the mutant (Fig. 27). Retransformation of *abrB*::TnYLB-1 fragment from the *abrB* mutant into the wild type FZB42 was also done in order to confirm that the mutation was in the *abrB* gene. The PCR product of retransformation indicated that *abrB* gene containing transposon insertion has replaced the intact *abrB* gene in the wild type (Fig. 27).

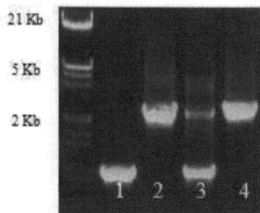

Figure 27. PCR product of *abrB* gene. Wild type FZB42 (lane1), *abrB* mutant (lane 2), complementation of *abrB* (lane 3) and retransformation of *abrB* (lane 4).

3.5.2.2 Effect of *abrB* mutation on growth of *L. minor* and *A. thaliana*

Bacterial culture of FZB42 wild type, *abrB* mutant and its complementation and retransformation were applied to Lemna biotest system in order to compare their effects on the growth of *L. minor* (Fig. 28 and Fig. 29). *L. minor* treated with the wild type bacterial culture revealed significant increase ($p \leq 0.05$) in growth parameter indicated by dry weight compared to control, mutant and retransformation. As predicted, complementation of the *abrB* mutant showed similar plant growth promoting activity to wild type FZB42, however it showed no significance difference compared to retransformation. The growth of plants treated with complementation cells increased significantly ($p \leq 0.05$) compared to control and mutant.

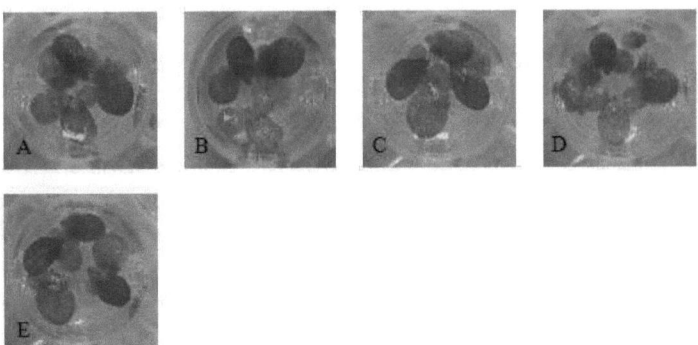

Figure 28. Influence of *abrB* mutation on plant growth promoting ability of *B. amyloliquefaciens* FZB42 on *L. minor*. The figure shows the representative plant response to the bacterial treatment at 10 days. A. *B. amyloliquefaciens* FZB42, B. Control, C. *abrB* mutant, D. Complementation of *abrB*, E. Retransformation of *abrB*.

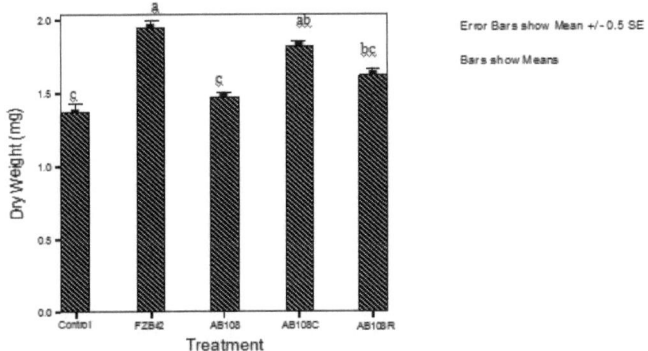

Figure 29. Growth stimulating effects of *abrB* mutation on *L. minor*. FZB42 : *B. amyloliquefaciens* FZB42, AB108 : *abrB* mutant, AB108C : Complementation, AB108R : Retransformation. Values shown represent the mean of fresh weight ± standard error. Different letters indicate means that differ significantly ($P \leq 0.05$).

The effect of the bacterial culture of the *abrB* mutant towards the growth of *A. thaliana* was shown in Fig. 30 and Fig 31. Immersion of the roots of *A. thaliana* into the bacterial culture of FZB42 improved the growth of the plant significantly ($p \leq 0.05$) when compared to the control and the retransformation. Application of the mutant slightly reduced the fresh weight of *A. thaliana* compared to FZB42. The treatment with complementation cells showed no significant differences compared to the control, mutant and retransformation altough the fresh weight of *A. thaliana* in this treatment showed an increase tendency.

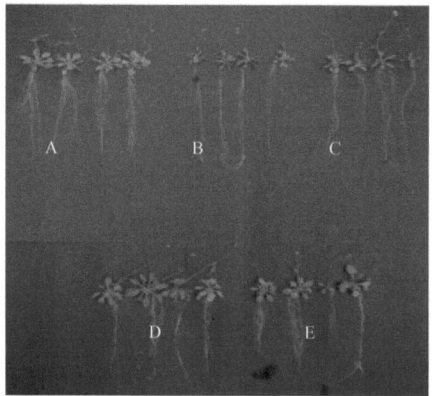

Figure 30. Influence of *abrB* mutation on plant growth promoting ability of *B. amyloliquefaciens* FZB42 on *A. thaliana*
The figure shows the representative of plant growth to the bacterial treatment at 21 days. A. *B. amyloliquefaciens* FZB42, B. Control, C. *abrB* mutant, D. Complementation of *abrB*, E. Retransformation of *abrB*.

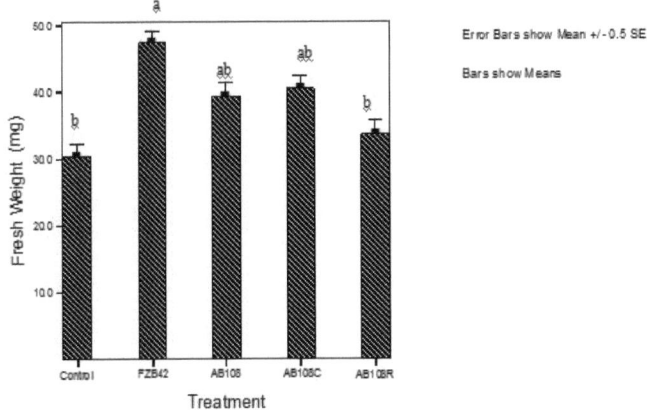

Figure 31. Growth stimulating effects of the *abrB* mutation on *A. thaliana*. FZB42 : *B. amyloliquefaciens* FZB42, AB108 : *abrB* mutant, AB108C : Complementation of *abrB*, AB108R : Retransformation of *abrB*. Values shown represent the mean of fresh weight ± standard error. Different letters indicate means that differ significantly (P ≤ 0.05).

3.5.3 *B. amyloliquefaciens* FZB42 *RBAM_017410*::TnYLB-1

The *RBAM_017410* is a small gene of 185 bp where its function is still unknown. However, it is homologous (76%) to ribonucleoside-diphosphate reductase small subunit in *B. subtilis*. The region of *RBAM_017410* inserted with TnYLB-1 and its flanking region is shown Fig. 32.

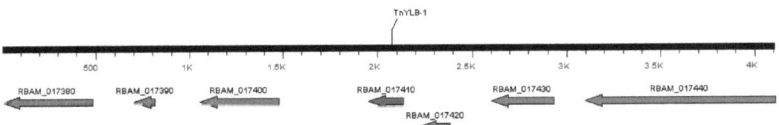

Figure 32. Genomic organization of *RBAM_017410* region carrying the TnYLB-1 insertion and its flanking regions.

3.5.3.1 Complementation of *RBAM_017410* mutant

To complement the RBAM_017410 mutant, the *RBAM_017410* gene (186 bp) plus 261 bp upstream region and 123 downstream region was amplified using 410-up-SacII and 410-dw-Eco91I primers, containing SacII and Eco91I restriction site, respectively. The resulting fragment (586 bp) was cloned into linearized *SacII/Eco91I* pUC18 plasmid which had an *Amy* cassette (pVBF) (Fig. 33). The ligated DNA was transformed into CaCl$_2$-competent *DH5α* cells. To verify the insertion site, the plasmid DNA extracted from *DH5α* was amplified using 410-up-SacII and 410-dw-Eco91I primers. The plasmid with correct insertion fragment was then transformed into *RBAM_017410* mutant.

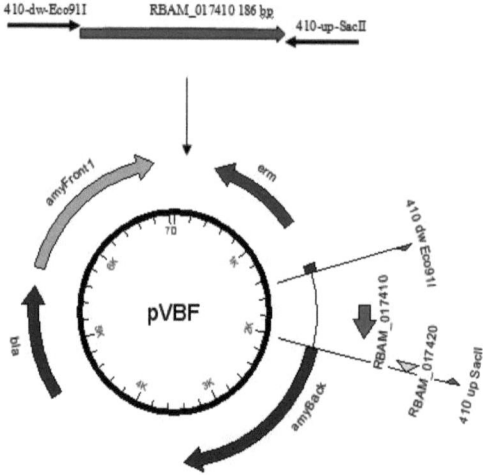

Figure 33. Strategy for construction of pUC18-Δ *RBAM_017410* cassette.

Amplification of FZB42 chromosomal DNA by PCR using primer 410-up-SacII and 410-dw-Eco91I which amplified *RBAM_017410* gene showed a fragment around 500 bp length, whereas the mutant revealed a fragment 2000 bp in length, indicating that there was insertion of the TnYLB-1 transposon (Fig. 34). Complementation to replace the disrupted *RBAM_017410* gene was done by inserting the intact gene from the wild type. Two fragments were obtained in complementation when its chromosomal DNA was amplified with primer 410-dw-Eco91I and 410-up-SacII, suggesting the intact gene has been successfully inserted. Retransformation was also done to verify that the *RBAM_017410* gene was interrupted by transposon insertion. This was done by transforming chromosomal DNA mutant to the FZB42 wild type. Amplifying the chromosomal DNA of retransformation showed that the gene from mutant has been transformed into wild type (Fig. 34).

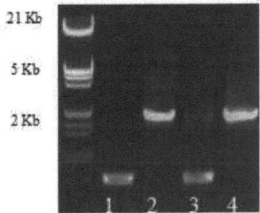

Figure 34. PCR product of *RBAM_017410* gene. Wild type FZB42 (lane1), *RBAM_017410* mutant (lane 2), complementation of *RBAM_017410* (lane 3) and retransformation of *RBAM_017410* (lane 4).

3.5.3.2 Effect of RBAM_017410 mutation on growth of *L. minor* and *A. thaliana*

In the preliminary screening, the *RBAM_017410* mutant showed reduction in plant growth promoting activity. Application of bacterial culture of the mutant reduced significantly (p ≤ 0.05) the growth of *L. minor* compared to wild type. The complementation of *RBAM_017410* significantly improved (p ≤ 0.05) the dry weight of the plants compared to control and mutant, however it showed no significant difference to the retransformation. Bacterial culture of FZB42 revealed significant (p ≤ 0.05) increase compare to all treatment in Lemna biotes system (Fig. 35 and 36).

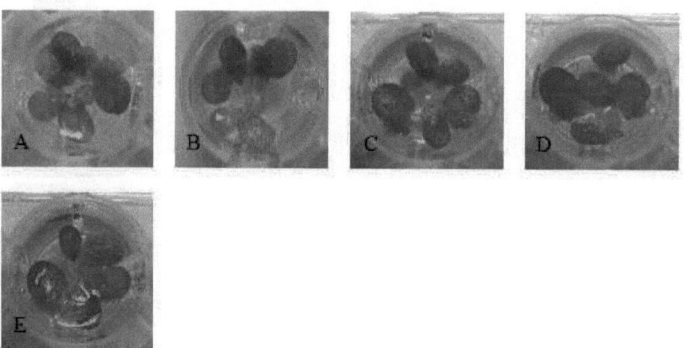

Figure 35. Influence of *RBAM_017410* mutation on plant growth promoting ability *B. amyloliquefaciens* FZB42 on *L. minor*. The figure shows the representative plant response to the bacterial treatment at 10 days. A. *B. amyloliquefaciens* FZB42, B. Control, C. *RBAM_017410* mutant, D. Complementation of *RBAM_017410* , E. Retransformation of *RBAM_017410*.

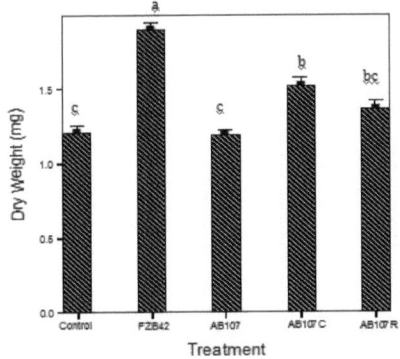

Figure 36. Growth stimulating effects of *RBAM_017410* mutation on *L. minor*. FZB42 : *B. amyloliquefaciens* FZB42, AB107 : *RBAM_017410* mutant, AB107C : Complementation of *RBAM_017410*, AB107R : Retransformation of *RBAM_017410*. Values shown represent the mean of dry weight ± standard error. Different letters indicate means that differ significantly (P ≤ 0.05).

Immersion the roots of *A. thaliana* in bacterial cultures of the *RBAM_017410* mutant reduced the growth of the plant significantly (p ≤ 0.05) compared to wild type. The application of the complementation revealed no significant increase of plant fresh weight compared to control and mutant, but showed significance increase (p ≤ 0.05) compared to retransformation. The treatment of wild type and complementation showed no significant differences in plant growth promotion capability (Fig. 37 and 38).

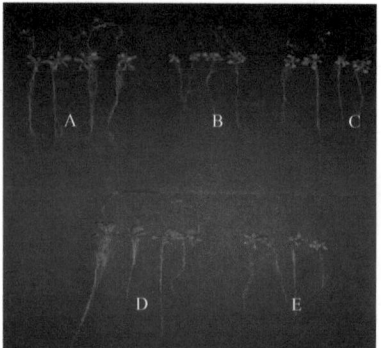

Figure 37. Influence of RBAM-017410 mutation on plant growth promoting ability of *B. amyloliquefaciens* FZB42 on *A. thaliana*. The figure shows the representative of plant growth to the bacterial treatment at 21 days. A. *B. amyloliquefaciens* FZB42, B. Control, C. *RBAM_017410* mutant, D. Complementation of *RBAM_017410* , E. Retransformation of *RBAM_017410*.

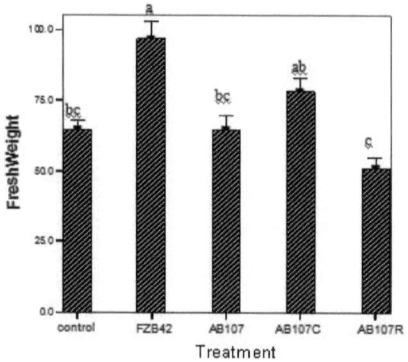

Figure 38. Growth stimulating effects of *RBAM_017410* mutation on *A. thaliana*. FZB42 : *B. amyloliquefaciens* FZB42, AB107: RBAM_017410 mutant, AB107C: Complementation of *RBAM_017410*, AB107R: Retransformation of *RBAM_017410*. Values shown represent the mean of dry weight ± standard error. Different letters indicate means that differ significantly ($P \leq 0.05$).

3.6 Colonization of *B. amyloliquefaciens* FZB42 and its mutants in *A. thaliana* roots growing in gnotobiotic system

It is generally assumed that colonization of plant tissue is a crucial step in plant-bacteria interactions. Here I showed the ability of *B. amyloliquefaciens* FZB42 and its mutants to colonize the roots of *A. thaliana* in a gnotobiotic system using a scanning electron microscope (SEM) and a confocal laser scanning microscope (CLSM). Labeling of *B. amyloliquefaciens* FZB42 with GFP protein was performed via homologous recombination (Ben *et al.* 2011). The GFP-tagged chromosomal DNA of FZB42 was then used to transfer the GFP into the mutants generating GFP-labeled mutants. The colonization of FZB42 and its mutants tagged with GFP were studied using CLSM in *A. thaliana* roots.

After seven days of incubation in growth chamber, colonization of *B. amyloliquefaciens* FZB42 and its mutants on the roots of *A. thaliana* was observed under SEM and CLSM. The results from CLSM showed that the wild type completely colonized the whole root surface including the root tip and the shed border cells (Fig. 39 A), the growing side root (Fig. 39 B) as well as the junctions between epidermal cells and root hairs (Fig. 39 C). Further observation by SEM revealed morphology of the wild type which formed an extracellular polymeric matrix around the bacterial cells (Fig. 40), which seemed to involve in adhering bacteria to the surface of the roots, besides it enabled the bacteria to form multicelulllar aggregation on the roots. The shape of the wild type cell appeared as a compact rod, covered with polymeric matrix when colonizing the root.

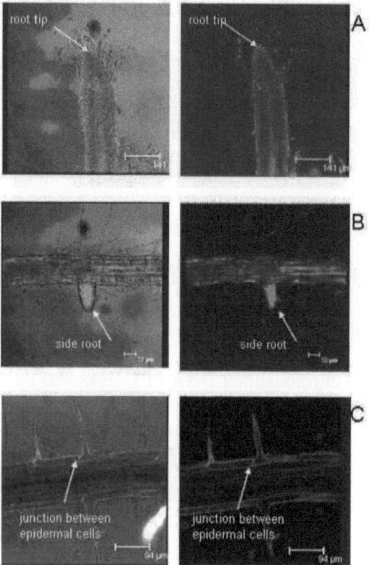

Figure 39. CLSM image of *B. amyloliquefaciens* FZB42 on *A. thaliana* root: 7 days after inoculation. Left images show an overlay for orientation. A) Root tip of the main root; FZB42 colonizes the whole root surface including the root tip as well as the border cells. B) Growing side root; FZB42 colonizes the side root and the whole root surface; C) Epidermal cells with root hairs; FZB42 colonizes in junctions between epidermal cells and on root hairs (by courtesy of Barbara Beator).

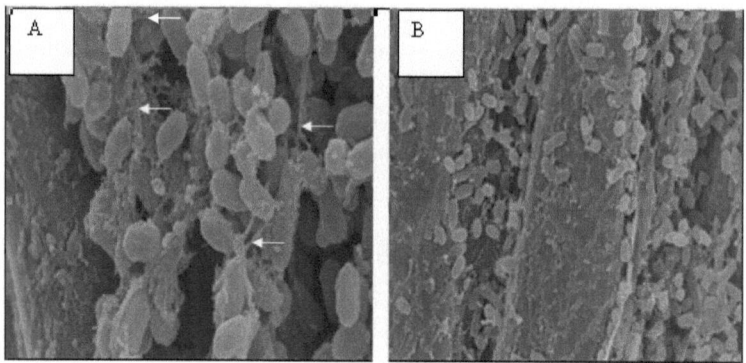

Figure 40. SEM of *B. amyloliquefaciens* FZB42 colonizing *A. thaliana* roots. Note the polymeric matrix surround bacterial cells as indicated by the arrows A. Cells with 25.00K magnification, B. Cells with 10.00K magnification.

Contrary to the wild type which colonized the whole root surface, all mutants impaired their ability to colonize the root of *A. thaliana*. Two biofilm mutants used in this experiment (*yusV* and *degU* mutants) showed ability to colonize only in some part of the root. The *yusV* mutant was found on the border cells but not in the root tip, surface of the root or side root (Fig. 41 A). However, there was a clump of bacterial cells opposite the cells (Fig. 41 B). This clump might originate from colonized border cells which detached during root preparation. The cells of *yusV* mutants showed significantly reduction in extracellular polymeric matrix production and most of the cells were not encased by this matrix as in FZB42 (Fig. 42 A). There was a clear shape difference between the FZB42 and *yusV* mutant, where the mutant appeared slender rod instead of compact rod.

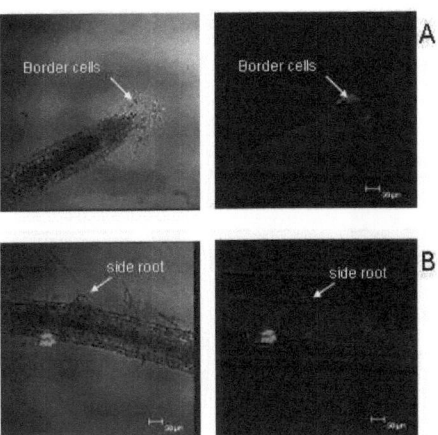

Figure 41. CLSM image of *yusV* mutant on *A. thaliana* roots: 7 days after inoculation. Left images show an overlay for orientation. A) Root tip; The root surface and the root tip are not colonized. *yusV* mutant colonizes the border cells. B) Side root; The side root and the root surface are not colonized (by courtesy of Barbara Beator).

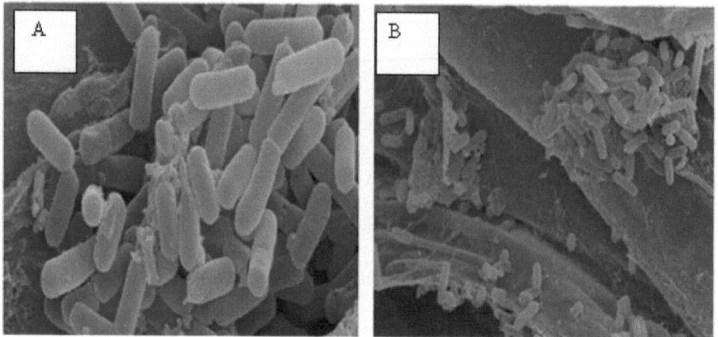

Figure 42. SEM of *yusV* mutant colonizing *A. thaliana* roots. A. Cells with 25.00K magnification, B. Cells with 10.00K magnification.

As for the *yusV* mutant, the biofilm *degU* mutant showed similar root colonization, where it was found in small amount in some border cells but not in the root tip, the root surface or side root (Fig. 43 A-B). The shape of the *degU* mutant cell appeared as slim rods when colonizing the root and the polymeric matrix connected bacterial cells (Fig. 44 A-B).

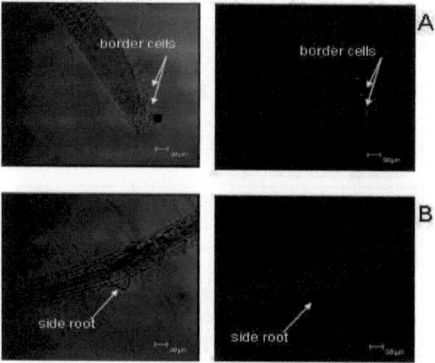

Figure 43. CLSM image of *degU* mutant on *A. thaliana* roots: 7 days after inoculation. Left images show an overlay for orientation. A) Root tip; The root surface and root tip are not colonized but *degU* mutant colonizes the border cells. B) Side root; The root surface and the side root are not colonized (by courtesy of Barbara Beator).

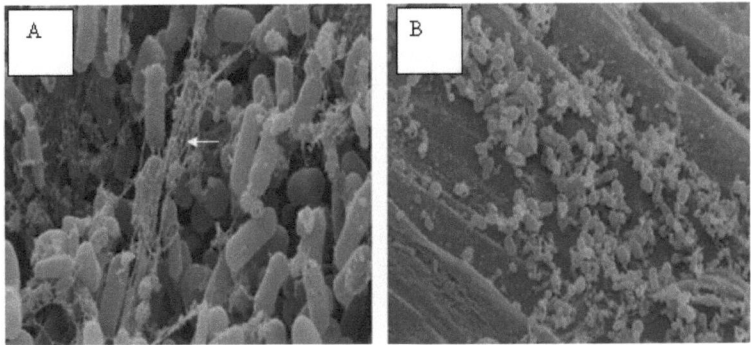

Figure 44. SEM of *degU* mutant colonizing *A. thaliana* roots. Note the polymeric matrix connecting bacterial cells as indicated by the arrow A. Cells with 25.00K magnification, B. Cells with 10.00K magnification.

Unlike the wild-type strain, the PGPR *nfrA* mutant failed to colonize the entire root surface (Fig. 45 A-B). It only colonized a small region of the roots that were root surface and border cells. SEM of *nfrA* mutant colonizing plant root showed reduction of extracellular polymeric matrix formation. The shape of the cells was slender rod and there were many dead bacteria cells among the colony (Fig. 46 A-B).

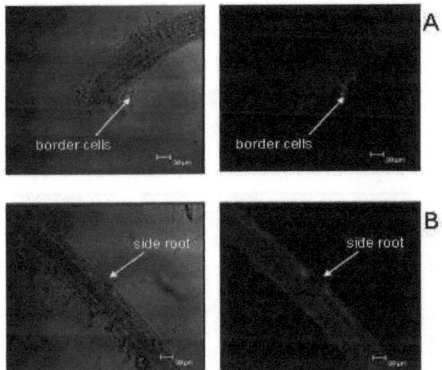

Figure 45. CLSM image of *nrfA* mutant on *A. thaliana* roots: 7 days after inoculation. Left images show an overlay for orientation. A) Root tip; The root surface and root tip are not colonized but *nfrA* mutant colonizes the border cells. B) Side root; The root surface is colonized but not the side root (by courtesy of Barbara Beator).

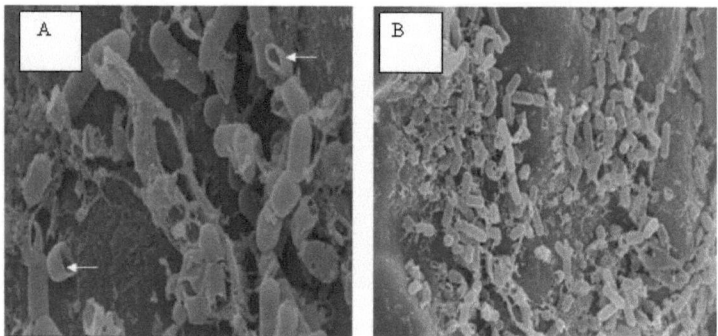

Figure 46. SEM of *nfrA* mutant colonizing *A. thaliana* roots. Note the dead bacterial cells as indicated by the arrows A. Cells with 25.00K magnification, B. Cells with 10.00K magnification.

Only few and small regions of the roots were colonized by the PGPR *abrB* mutant and significantly reduced extracellular polymeric matrix formation was observed compared with the wild type. Colonization was only detected in the border cells but not in the root tip or the root surface next to the tip (Fig. 47). The shape of the *abrB* mutant cell when colonizing the root was slender rod and many empty cells appeared in the colony (Fig. 48).

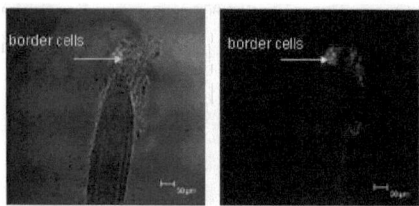

Figure 47. CLSM image of *abrB* mutant on *A. thaliana* roots: 7 days after inoculation. The left image shows an overlay for orientation. Root tip; The root surface and root tip are not colonized but *abrB* mutant colonizes the border cells (by courtesy of Barbara Beator).

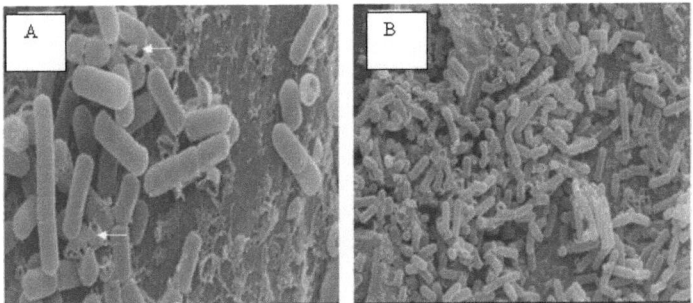

Figure 48. SEM of *abrB* mutant colonizing *A. thaliana* roots. Note the dead bacterial cells as indicated by the arrows A. Cells with 25.00K magnification, B. Cells with 10.00K magnification.

The PGPR *RBAM_017410* mutant was not able to be seen with CLSM when colonizing the root of *A. thaliana* (Fig. 49). This might be due to low emission of the GFP protein in this mutant. However, I could characterize the *RBAM_017410* mutant by SEM while colonizing the root showing the slender rod shape of the cell with reduction of extracellular polymeric matrix production (Fig. 50).

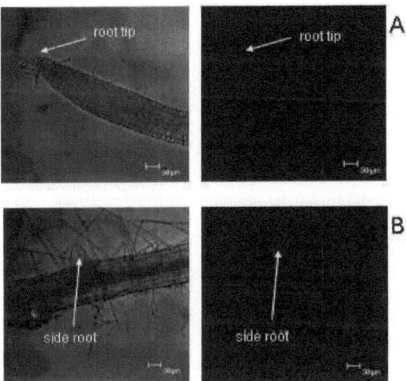

Figure 49. CLSM image of *RBAM_017410* mutant on *A. thaliana* roots: 7 days after inoculation. Left images show an overlay for orientation. A) Root tip; The root surface, root tip and border cells are not colonized. B) Side root; The root surface and the side root are not colonized (by courtesy of Barbara Beator).

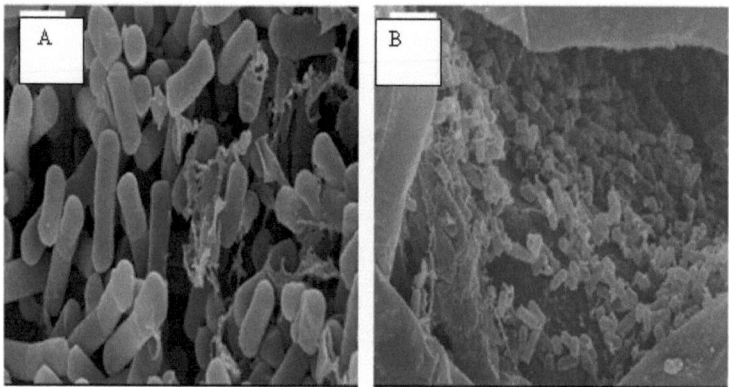

Figure 50. SEM of *RBAM_017410* mutant colonizing *A. thaliana* roots. A. Cells with 25.00K magnification, B. Cells with 10.00K magnification.

The interesting finding from this colonization experiment was a distinct mode of colonization between the wild type and the mutants. Whilst the wild type colonized the whole root surface including the root tip and the border cells as well as the growing side root, the mutants only colonized small region of the root. When comparing the cell shape of the wild type and the mutants colonizing the root of *A. thaliana*, clear differences became visible. The shape of the wild type cell was dumpy rod and extracellular polymeric matrix encased cells (Fig. 40), whereas the cell morphology of all the mutants were mostly slender rod with reduced extracellular polymeric matrix formation. There was not much difference in the cell shape among mutants, except that in *nfrA* mutant and *abrB* mutant many empty cells appeared (Fig. 46 and 48).

3.7 MALDI-TOF MS analysis of metabolites released by *B. amyloliquefaciens* FZB42 in plant-bacteria interactions

In order to determine which metabolites released by *B. amyloliquefaciens* FZB42 when interacted with *L. minor*, I decided to detect the metabolites released by mass spectrometric analysis. MALDI-TOF MS was carried out in Steinberg medium and *L. minor* inoculated with *B. amyloliquefaciens* FZB42 and the mutants (incubated for seven days at 22°C). In this analysis I found production of surfactin in the Steinberg medium and in the *L. minor* when inoculated with wild type and two other mutants (*degU* and auxotroph mutant). Surfactin is a biosurfactant which displays an array of amazing activities, such as hemolytic, hypocholesterolemic agent, antitumoral, antimicrobial and antiviral. The biological role of surfactin is thought as supporting colonization of surfaces and acquisition of nutrients through their surface-wetting and detergent properties (Peypoux *et al.*1999).

Analysis of the Steinberg medium containing *L. minor* and wild type showed the peaks that corresponded to surfactin production (m/z 1046.8, 1060.8 and 1074.8) (Fig. 51 A). Detection of *L. minor* extract grown in Steinberg medium inoculated with wild type revealed production of the similar metabolite (Fig. 51 B). Analysis of the Steinberg medium and the *L. minor* extract inoculated with auxotroph mutant indicated that surfactin was also produced in this treatment (Fig. 51 C-D). Inoculation of the *degU* mutant into the Steinberg medium planted with *L. minor* showed that Steinberg medium as well as *L. minor* extract also contained surfactin even though in the latter only two peaks of surfactin were detected (m/z 1060.8 and 1074.8) (Fig. 51 E-F).

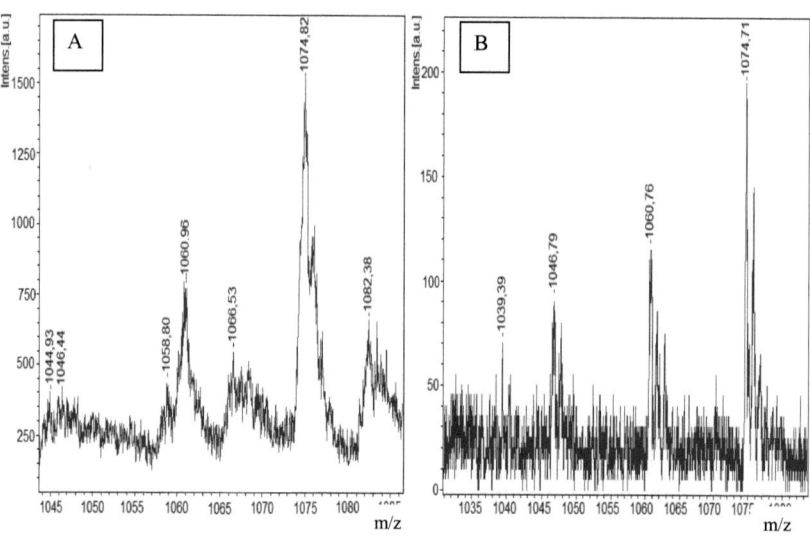
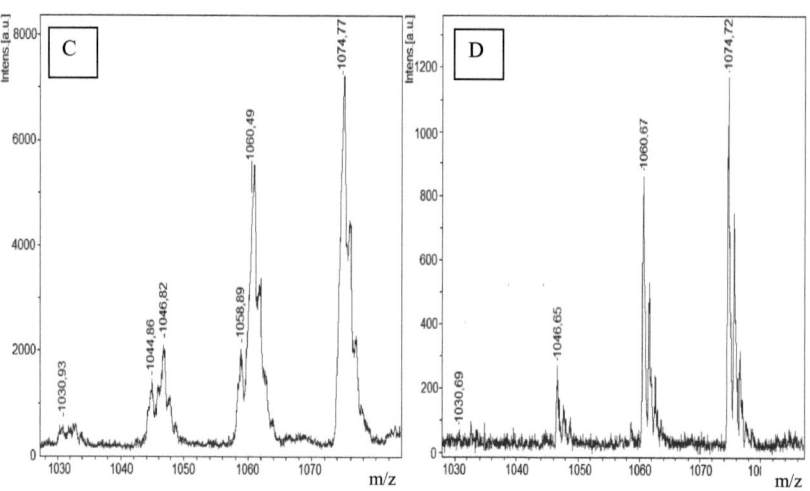

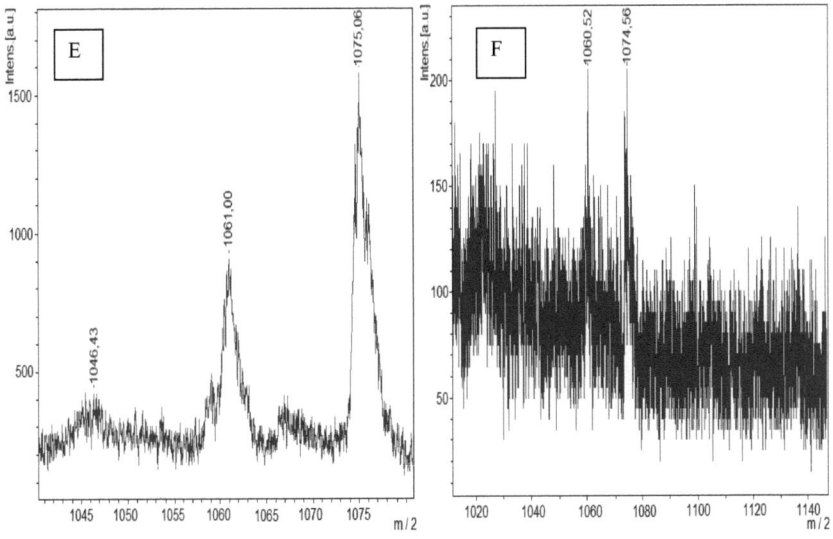

Figure 51. MALDI-TOF MS analysis of surfactin produced by *B. amyloliquefaciens* FZB42 and its mutants. A-B. surfactin detected in Steinberg medium and in *L. minor* inoculated with wild type. C-D surfactin detected in Steinberg medium and in *L. minor* inoculated with auxotroph mutant. E-F. Surfactin detected in Steinberg medium and in *L. minor* inoculated with degU mutant.

3.8 Screening for antibiotic mutants

*B. amyloliquefacien*s FZB42 is a potent producer of polyketides, such as bacillaene, difficidin, and macrolactin, as well as cyclic lipopeptides, such as surfactins, fengycins, and Bacillomycin D. These compounds are biosynthesized in a 4'-phosphopantetheine transferase (Sfp)-dependent fashion, while the production of the antibacterial dipeptide bacilysin is independent of Sfp (Koumoutsi *et al.* 2004; Chen *et al.* 2006; Chen *et al.* 2007; and Chen *et al.* 2009c). From the genome analysis of *B. amyloliquefaciens* FZB42 there was no evidence for ribosomally produced antibacterial substances such as lantibiotics and other bacteriocins (Chen *et al.* 2007), however, Butcher and Helmann (2006) showed that the *sfp* mutant CH3, that can not produce polyketides and lipopeptides, produced at least one substance that was effective against *Bacillus subtilis* CU1065 and particularly against its *sigW* mutant HB0042.

Using tranposon TnYLB-1, I tried to investigate the capacity of *B. amyloliquefaciens* FZB42 for the production of secondary metabolites of ribosomal origin. For this purpose, I have used the *sfp* mutant strains CH5 which lost the ability to produce all nonribosomally formed metabolites of FZB42 that were synthesized according to the thiotemplate mechanism by knocking-out the *sfp*-4'-phosphopantetheine transferase gene responsible for cofactor loading to the thiotemplate reaction centers (Chen, 2009). The *B. amyloliquefaciens* FZB42 mutant strain CH5 was transferred with transposon TnYLB-1 using the same method as for wild type transformation. All mutants were then screened in subsequent spot-on-lawn tests on *Bacillus subtilis* HB0042 for the loss of antibacterial activity.

In this screening, several mutants were found unable to produce antibacterial compounds. One of the mutants, WY01 (Fig 52), was selected and studied further. Sequencing the WY01 mutant revealed a gene cluster around *RBAM_029230* with hypothetical proteins. Based on MALDI-TOF mass spectrometry, the product of this gene cluster was bacteriocin. Extensive blast of known bacteriocin genes against the cluster uncovered the similarity to the circular bacteriocin uberolysin. For this reason it was assumed that the product of the gene cluster around *RBAM_029230* is a circular peptide, and this is the antibacterial substance against *Bacillus subtilis*. This new antibacterial peptide is named "amylocyclin A" (Scholz, 2011). The insertion of TnYLB-

1 transposon in the *degU* gene of the CH5 strain (RSpMarA2) produced a substance which was identified as a thiazole/oxazole-modified microcin (TOMM). This novel metabolite was also a ribosomally synthesized antibacterial substance and named as "plantazolicin" (Scholz *et al.* 2011). Both new antibiotics are narrow spectrum antibacterial compounds which have antagonistic activity towards closely related Gram-positive *bacilli*.

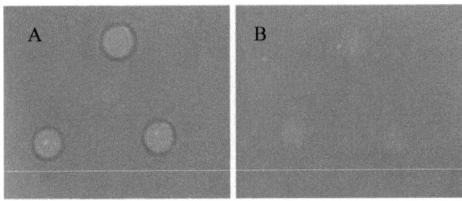

Figure 52. Spot on lawn test of WY01 mutant . A. CH5, B. WY01 mutant. Note clear zone was lost in the mutant (By courtesy of Zhiyuan Wang).

3.9 Screening for nematocidal mutants

Many bacteria such as *Pseudomonas aeruginosa*, *P. fluorescens* CHA0, *Bacillus thuringiensis* and *Paenibacillus macerans* (Ali *et al.* 2002; Siddiqui *et al.* 2005; Wei *et al.* 2003; and Oliveira *et al.* 2009) have the ability to produce active nematocidal metabolites against nematode pests. Here the nematocidal activity of *B. amyloliquefaciens* FZB42 was tested and the mutants were screened to discover the genes responsible for nematodical production. Transposon TnYLB-1 mutagenesis was used to generate the mutants of *B. amyloliquefaciens* FZB42 and the transposant were then screened to find the mutants that impaired nematocidal activity.

The screening was performed in China (Quisheng He – Yunan University, Kunming - China), four mutants were found which had reduction of nematocidal activity compared to FZB42 (Table 7). In this screening the survival rate of the nematode *Caenorhabditis elegans* was counted every four, eight and twenty hours and then compared with the wild type. Blast analysis of the DNA sequence form the mutants showed that the transposon disrupted the *prkA, yhdY, RBAM_007470, ywmC* and *mlnD* genes.

Table 7. *In vitro* effects of *B. amyloliquefaciens* FZB42 and its mutants on *C. elegans* livability

Strain[a] \ Hours	FZB42	E23	F5	F21	F35
4h	70.42%±8.47%	80.64%±6.18%	77.12%±6.43%	70.63±7.84%	66.64±10.82%
8h	41.71%±7.55%	56.83%±6.5%	54.89%±6.59%	44.48±9.67%	43.88±11.86%
20h	7.02%±3.86%	15.99%±7.66%	12.9%±7.72%	6.8±3.51%	7.6±4.8%

[a]Strain FZB42 (wild type), E23 (*ywmC*), F5 (*prkA,yhdY, RBAM_007470*), F21 (*RBAM_007470*), and F35 (*ywmC* and *mlnD*) (By courtesy of Quisheng He).

4. Discussion

As a potent plant growth-promoting rhizobacteria, the cellular functions of *B. amyloliquefaciens* FZB42 are important to be studied. Several modes of action such as phytase production, auxin synthesis and production of antibiotics which support its beneficial activity towards the growth of the plant have been elucidated (Idriss *et al.* 2002; Idriss *et al.* 2007; Koumutsi *et al.* 2004; and Chen *et al.* 2009b). However, not all of the molecular mechanisms are completely understood. In this work, I presented additional information regarding molecular mechanisms of plant growth promotion activity of *B. amyloliquefaciens* FZB42 generated by transposon mutagenesis.

4.1 *Himar1* transposon mutagenesis of *B. amyloliquefaciens* FZB42

Transposon mutagenesis is a powerful tool in molecular biology used to create random insertion mutagenesis. It has been applied to and established *in vitro* and *in vivo* in most scientifically and industrially relevant organisms ranging from prokaryote to eukaryote (Petzke and Luzhetskyy, 2009). Among many transposons used, *mariner*-based transposons gain much attention recent time. *Himar1* has been useful for *in vitro* and *in vivo* transposon mutagenesis of a variety of bacteria due to its advantage features in transposition which does not require host-specific factors and inserts more randomly into the chromosome, since the *mariner* transposon requires only the dinucleotide TA (Akerley *et al.* 1996; Ashour and Hondalus 2003; and Zhang *et al.* 2000). In the present work, I applied the *mariner* transposable element *Himar1* (TnYLB-1) transposon that has been previously used in *B. subtilis* (Le Breton *et al.* 2006) to create a transposon library in *B. amyloliquefaciens* FZB42.

Three different plasmids containing a mariner-based *Himar1* tranposon (pMarA, pMarB and pMarC) were used in transposon mutagenesis of *B. amyloliquefaciens* FZB42. Both pMarA and pMarB contained the *Himar1* transposase gene; however they have different promoters that drive the expression of transposase. The pMarA plasmid is under the control of the housekeeping σ factor σ^A of *B.subtilis*, whereas pMarB employs σ factor σ^B. The pMarC does not contain the transposase gene as well as its promoter. I have successfully transformed plasmids pMarA and pMarC into *B. amyloliquefaciens*

FZB42, but due to unknown reasons I failed to transform pMarB. However, this failure might be due to the different promoters used in the plasmid.

The frequency of transposition of TnYLB-1 in *B. amyloliquefaciens* FZB42 in this work (10^{-2}) is similar to the frequency of transposition in *B. subtilis*, indicating that TnYLB-1 is inserted in the high rate into the genome *B. amyloliquefaciens* FZB42 as in *B. subtilis*. Interestingly, the Kan^r clones of *B. amyloliquefaciens* FZB42 which represent the transposition events were recovered at an efficiency of 2.6 x 10^7 which is higher than in *B. subtilis* (6.2 x 10^6). In order to examine the utility of TnYLB-1 for mutant isolation and to prove that it is an efficient tool to generate transposon library in *B. amyloliquefaciens* FZB42, the temperature-resistant Kan^r clones were transferred to a glucose minimum medium to screen for auxotrophic mutations. From 778 Kan^r colonies, I observed seven mutants that have failed to grow under this condition that corresponded to 0.89% recovery of auxotrophs. In contrast, lower frequency of auxotrophy (0.15%) was obtained in a screen of *Francisella tularensis* transposon mutants using *Himar1* transposon on Chamberlain's chemically defined medium (Maier *et al.* 2006). However, similar frequency of auxotrophy of transposon mutants was reported in the screening of *Burkholderia pseudomallei* on M9 minimal-glucose kanamycin media (0.72%) and *Bacillus subtilis* on glucose-minimal media (1%) (Rholl *et al.* 2008 and Le Breton *et al.* 2006). Moreover, the result obtained here is comparable to the isolation of auxotrophs in other bacteria using other transposon system which often results in 1 to 2% recovery rate (Petzke and Luzhetskyy, 2009).

Further characterization of two out of seven auxotroph mutants by Southern Blot analysis and mapping of genomic insertion sites showed a single transposon insertion in the genome. The insertion sites of the two auxotroph mutants were located in the *pabB* and *hisJ* genes, respectively. These genes encode para-aminobenzoate synthase (subunit A) and histidinol phosphate phosphatase involved in folate and histidine biosynthetic pathways, respectively. The respective mutants are therefore para-aminobenzoate (*pabB*) and histidine (*hisJ*) auxotroph. These auxotrophies were experimentally confirmed since the growth of the *pabB* and *hisJ* mutants in glucose minimal medium was restored by the addition of para-aminobenzoic acid (PABA) and histidine, respectively. These results have demonstrated the utility of TnYLB-1 for transposon mutagenesis in *B. amyloliquefaciens* FZB42. After examining that the TnYLB-1 has been transposed in the genome of *B. amyloliquefaciens* FZB42, I began to implement transposon mutagenesis strategy to generate a transposon library for the desired phenotype.

4.2 Identification of genes involved in swarming motility and biofilm formation in *B. amyloliquefaciens* FZB42 genome

In order to detect genes that may be potentially involved in plant-associated lifestyle, especially in rizhosphere competence, I applied the TnYLB-1 transposon to a generate transposon library for screening the transposants with swarming and biofilm mutant phenotype. Swarming is the fastest known bacterial mode of surface translocation that enables the rapid colonization of a nutrient-rich environment and host tissues. This complex multicellular behavior requires the integration of chemical and physical signals, which leads to the physiological and morphological differentiation of the bacteria into swarmer cells (Verstraeten *et al.* 2009). They become multinucleate, elongate, synthesize large numbers of flagella, secrete surfactants and advance across the surface in coordinated packs (Berg, 2005). Bacteria capable of swarming motility produce highly organized communities initially consisting of vegetative cells, i.e. the swimmer cells, which undergo a co-ordinated surface induced differentiation process characterized by the production of hyperflagellated, elongated, multinucleate cells, i.e. the swarm cells (Senesi *et al.* 2002). Swarming motility as a flagellum-dependent behavior has been implicated in biofilm formation, including colonization and bacterial virulence (Kirov *et al.* 2002; and Merritt *et al.* 2007). In *Pseudomonas aeruginosa*, *Salmonella typhimurium* and *Proteus mirabilis*, swarming ability as a part of motility behavior is associated with their pathogenesis due to the level of specific virulence factors production that is higher during their swarm-cell state (Allison *et al.* 1994; Wang *et al.* 2004; and Yeung *et al.* 2009). In addition, swarmer cells of *P. aeruginosa* also exhibits elevated adaptive antibiotic resistance to ciprofloxacin, gentamicin and polymyxin B, thus swarming is part of an alternative growth state and is a complex adaptation process that occurs in response to specific stimuli (Overhage *et al.* 2008).

There are many genetic determinants that govern the swarming motility. In *P. aeruginosa* the swarming associated genes functioned not only in flagellar or type IV pilus biosynthesis but also in transport, secretion and metabolism process. As much as 35 transcriptional regulators were found to be involved in swarming, including a variety of two-component sensors and response regulators (Yeung *et al.* 2009). In my screening for swarming mutant, one swarming-deficient mutant with transposon insertion was found in the *degU* gene. Interestingly, this mutant also showed impairing in biofilm production. This result is in contrast to the finding of Mireles *et al.* (2001) where a swarming mutant of *Salmonella enterica* defective in lipopolysaccharide (LPS) synthesis promotes biofilm formation. In that research, they found that O-antigen mutants of *S. enterica*, which are defective in swarming, were generally more proficient in biofilm formation than the wild type. The O-antigen has been postulated to provide a surfactant or wettability function that allows spreading of the swarmer colony. Surfactin, a cyclic lipopeptide from *B. subtilis* which rescued the swarming defect of the LPS mutants, inhibited biofilm formation. They have concluded that the absence of surfactants inhibits swarming but promotes biofilm formation and vice versa.

The *degU* gene which is involved in two-component response regulation has a key role in regulating several post-exponential phase processes in *B. subtilis*, including the activation and the inhibition of genetic competence, the inhibition of flagellar-based-motility, the activation of degradative enzyme production and activation of poly-γ-glutamic acid production (Verhamme *et al.* 2007). As swarming motility is a flagella-dependent behaviour, the inhibition of the genes required for the assembly of the flagella will also affect the swarming behaviour as shown in *B. amyloliquefaciens* FZB42 *degU* mutant. The *degU* mutant and retransformation of the *degU* showed that both mutants failed to colonize the 'swarm plates' (LB medium with 0.7% agar) after 24 hours incubation, whereas the wild type and complementation of *degU* colonized the entire swarm plates (Fig. 8). Amati *et al.* 2004 have reported that phosphorylated DegU (DegU~P) acts as a repressor of transcription of the *fla-che* operon which comprises the majority of the genes whose products are structural components of the flagellum, in addition to several genes involved in chemotaxis. Hence, repression of *fla-che* operon prevents the expression of genes coding for the hook and basal body components of the flagellum and for the σ^D transcription factor. A repressive role of DegU~P on the *fla-che* operon is supported by the fact that genes of this operon were among those that were down-regulated by the overproduction of *degU* (Maeder *et al.* 2002). In *B. subtilis*, very low levels of DegU~P activate swarming, while high levels of DegU~P inhibit swarming, however, the genes required for swarming motility that are regulated by very low levels of DegU~P are currently unknown (Verhamme *et al.* 2007).

Biofilms are structured communities of cells that are adherent to a surface, an interface, or each other are encased in a self-produced polymeric matrix (Stanley *et al.* 2003). Initiation of biofilm formation is characterized by interaction of cells with a surface or an interface as well as with each other. Once enough cells have aggregated, the biofilm begins to mature through the production of an extracellular matrix, which contributes greatly to the final architecture of the community (Branda *et al.* 2005). The biofilm offers many advantages for the bacteria. The first, bacteria experience a certain degree of shelter and homeostatis when residing within biofilm and one of the key components of this microniche is the surrounding extrapolymeric substance matrix. This matrix is composed of a mixture of components, such as exopolysaccharide (EPS), protein, nucleic acids and other substances. The EPS provides protection from a variety of environmental stresses, such as UV radiation, pH shift, osmotic shock, desiccation and also antibiotic (Davey and O'toole 2000; and Fleming 1993). The second, highly permeable water channels interspersed throughout the biofilm in the areas surrounding microcolonies gives an effective means of exchanging nutrients and metabolites with the bulk aqueous phase, enhancing nutrient availability as well as removal of potentially toxic metabolites (Costerton *et al.* 1995). And lastly, biofilm is also a place where gene transfer takes place. Hence, it helps acquisition of new genetic traits (Davey and O'toole, 2000).

In general, there are four universal systems routinely used to study the biofilm formation. Depending on the aims of the experiment, these system can be roughly categorized as flow cells biofilm which are usually analyzed using CSLM, 'Microtiter' biofilms which are formed on the surfaces of microtiter dish wells under standing culture conditions, Floating biofilms or pellicles which are formed at the air–liquid interface of a standing culture and finally, the bacterial colonies which grow on the surface of agar-solidified media (Branda et al. 2005). In my biofilm screening, I have applied screening of floating or pellicles formation in standing culture. Three biofilm mutants that were found in this screening showed defects in biofilm formation compared with the wild type. These mutants were unable to build wrinkled-upper layer of pellicles in the MSgg medium.

The biofilm *degU* mutant was impaired in its ability to form robust pellicles when incubated in screening medium. This mutant formed a thin membrane in the upper layer of the medium with little granules located in the center of surface (Fig. 9 B). The complementation using intact *degU* gene from the wild type restored the ability of mutant to form robust pellicle (Fig. 9 C) confirming the role of degU gene in formation of biofilm. The response regulator DegU controls the production of exoenzymes such as proteases, levan-sucrase, and α-amylase and it is also involved in competence development and motility. DegU is the second element of a two-component signaling system that is phosphorylated by the first component, DegS (Amati et al. 2004). The finding that the *degU* gene is involved in biofilm formation is in accordance with the result of Kobayashi (2007) and Verhamme *et al.* (2007) who investigated the role of *degU* in pellicle formation. In *B. subtilis* a low level of phosphorylated *degU* (DegU~P) is needed to activate the *yvcA* gene, coding for a membrane-bound lipoprotein, which is required for complex colony architecture. How *yvcA* functions to control this process is still unclear, although it is possible that it could form a structural component of the extracellular matrix or could be involved in a signaling process. However, the *yvcA* gene is not present in the genome of *B. amyloliquefaciens FZB42* (Verhamme *et al.* 2007). Kobayashi (2007) found the *yuaB* gene which had a role in pellicle formation under regulation of *degU*. It is required for a later stage of pellicle formation, such as floating of cell clusters to the surface of the medium. The *degU* mutation abolishes both flagellum and pellicle formation, suggesting that regulation of DegU activity may be a key to the transition between the two cell states.

The phenotype of biofilm *yusV* and *pabB* mutants showed almost the same appearances where both mutants formed flat and thin pellicles on the surface of MSgg media (Fig. 13 B and 17 B). These two mutants were not able to form robust and wrinkled pellicles as the wild type. Insertion of the intact *yusV* and *pabB* genes from the wild type restored pellicle formation in each mutant (Fig. 13 C and 17 C), confirming that both genes might be essential in biofilm formation. The function of the *yusV* gene is still unknown, however, it shows similarity to iron (III) dicitrate transport permease. In addition, the *yusV* nucleotide sequence shows 97% similarity to iron (III)-siderophore transporter (ATP binding component) of *Bacillus amyloliquefaciens* DSM7. The production of low-molecular-weight Fe (III) siderophores enables microorganisms to efficiently scavenge iron even in aerobic environments where iron exists primarily as insoluble hydroxides (Wandersman and Delepelaire, 2004). As iron is an essential nutrient for nearly all living organisms, the transport systems for the uptake of iron and iron-chelation complexes are therefore critical for growth (Ollinger *et al.* 2006). Hence, disruption of transposon TnYLB-1 in *yusV* gene which caused defect in pellicles formation of *B. amyloliquefaciens* FZB42 might be due to the growth defects. However, the precise explanation how this gene is involved in pellicle formation is still unclear.

The *pabB* gene encodes for para-aminobenzoate (PABA) synthase. In *E. coli* PABA is made from chorismate in two steps. First, *pabA* and *pabB* interact to catalyze transfer of the amide nitrogen of glutamine to chorismate, forming 4-amino-4-deoxychorismate (ADC). Association of *pabA* and *pabB* with one another to form the ADC synthase are commonly called PABA synthase. Second, the *PabC* protein then mediates elimination of pyruvate and aromatization of ADC to give PABA (Basset *et al.* 2003). PABA is a substrate of 7,8-dihydropteroate synthase (DHPS) in the de novo biosynthesis of folate which is important for the formation of purines, thymidylate, serine, methionine, glycine and formylmethionyl-tRNA (James *et al.* 2002). Disruption of *pabB* gene with TnYLB-1 transposon leads to inhibition of ADC synthase, as the ammonia that is generated from glutamine by *pabA* gene can not be used to aminate chorismate, producing ADC. Since there is no production of ADC, PABA will not be produced. Beside complementation of intact *pabB* gene which restored pellicles formation in this mutant, addition of 0.1 mM para-aminobenzoic acid into the MSgg medium also repaired pellicles formation (Fig. 17 E). Hence, the role of *pabB* gene in biofilm formation is indirect as it affects the growth of the mutant.

4.3 Identification of genes involved in plant growth promotion and colonization of the mutants in the root of *A. thaliana*

PGPR, which belong to diverse genera such as genera *Pseudomonas* and *Bacillus*, are rhizosphere-inhabiting bacteria that have drawn much attention in recent years because of their contribution to the biological control of plant pathogens and the improvement of plant growth (Persello-Cartieux *et al.* 2003). Several studies have already described the utility of Bacillus PGPR species for promoting plant growth. Determination of the mechanism of action of such bacteria revealed that they are able to solubilize phosphate (Idriss *et al.* 2002), produce plant hormone (IAA & gibberellins) (Idriss *et al.* 2007; and Chakraborty *et al.* 2005), siderophore and antifungal as well as antibacterial metabolite (Chen, 2009c; Joo *et al.* 2005; and Danielsson *et al.* 2006). Nonetheless, the mechanisms used by Bacillus spp. to stimulate plant growth are not fully understood.

In this study, I used a miniaturized *L. minor* biotest system which has been proven to compare the phytostimulatory effects exerted by FZB42 wild-type and mutant strains (Idriss *et al.* 2007). Duckweeds (Lemnaceae) possess physiological properties (small size, high multiplication rates, and vegetative propagation), which make them an ideal test system. Particularly interesting is the mechanism of propagation: under favourable environmental conditions both frond primordia grow out, thus forming new fronds by clonal propagation and producing a population of genetically homogeneous plants (Naumann *et al.* 2006). Hence *L. minor* is an attractive subject for investigating plant-microbe interactions (Lockhart *et al.* 1989). In this system, a selected clone from duck weed, *L. minor* ST, was cultured in Steinberg medium and it yielded reproducible results in plant growth test assays. By screening using the *L. minor* biotest system, I found three mutants which showed deficient in plant growth promotion activity. After sequencing and searching with BLAST, I identified that the TnYLB-1 disrupted *nfraA*, *abrB* and *RBAM_017410* genes. In order to validate, whether or not these genes were also responsible for the plant growth promotion phenotype in higher plant, I used *A. thaliana* to determine the effect of the mutants to its growth. The experiment with *A. thaliana* also offered a different system as in *L. minor* biotest system which used liquid media, whereas *A. thaliana* was grown in solid media.

Effective colonization of plant roots by PGPR plays an important role in growth promotion, irrespective of the mechanism of action (production of metabolites, production of antibiotics against pathogens, nutrient uptake effects, or induced plant resistance) (Bolwerk *et al.* 2003). Bacterial colonization in natural environment is mainly facilitated by biofilm formation. Bacterial biofilms established on plant roots could protect the colonization sites and act as a sink for the nutrients in the rhizosphere, hence reducing the availability of root exudate nutritional elements for pathogen stimulation or subsequent colonization on the root (Haggag and Timmusk, 2007). The ability to colonize plants is a multifactorial process requiring resistance to plant defence systems as well as the ability to initiate growth on plant surfaces, invade tissues and develop within the plant. In return for a safe and nutrient rich environment, the bacterium may provide the host with resistance to plant pathogens through the synthesis of antibiotics and enzymes and promote plant growth (Reva *et al.* 2004). As rhizosphere competence of biocontrol agents comprises effective root colonization combined with the ability to survive and proliferate along growing plant roots over a considerable time period, in this study I also examined the colonization of FZB42 and its mutants in *A. thaliana* roots using SEM and CLSM.

The PGPR *nfrA* mutants exhibited reduction in plant growth promoting activity either in Lemna system (26%) or *A. thaliana* (40%). Complementation of the gene restored the growth promoting activity, whereas retransformation from the mutant recuded the growth promoting effect (Fig. 21-24). In *B. subtilis*, *nfrA* gene is a putative essential oxidoreductase which is induced under heat shock and oxidative stress conditions. This gene belongs to the class III heat shock genes and its transcription is induced in a σ^D-dependent manner at the onset of the stationary phase and also by heat stress from a σ^A-dependent promoter overlapping the σ^D promoter. Sigma factor σ^D is usually necessary to transcribe genes involved in motility and cell division (Moch *et al.* 2000; and Moch *et al.* 1998). DNA microarray analysis revealed that *nfrA* is induced by superoxide stress and H_2O_2 stress. These results indicate that the putative essential NADPH-dependent oxidoreductase *nfrA* (*ywcH*) might play an important role in the oxidative stress response (Mostertz, *et al.* 2004). The *nfrA* was postulated to be an essential protein in *B. subtilis* but not in *S. aureus*. However, the enzyme could have a significant function in the bacterial stress response during phases of oxidative stress in infections caused by *S. aureus* (Streker *et al.* 2005). The precise role how *nfrA* gene involves in growth promoting activity is still unclear. However, Rudrappa *et al.* 2007 have proposed that oxidative stress results in down-regulating of biofilm formation of *B. subtilis* on *A. thaliana* roots. In his study, biofilm formation on *A. thaliana NahG* plants was suppressed by the presence of catechol on the root surface (and in the surrounding area), resulting in ROS (reactive oxygen species)-mediated downregulation of genes required for biofilm formation by *B. subtilis*. It has been demonstrated in rice and *Arabidopsis* systems that higher catechol levels result in the production of superoxides and H_2O_2 leading to increased ROS generation (Yang *et al.* 2004; and van Wees and Glazebrook, 2003). By SEM examination, a defect in biofilm formation and change in the shape of the cell was shown, when the *nfrA* mutant colonized *A. thaliana* roots (Fig. 46). Distribution of *nfrA* mutant cells on the *A. thaliana* roots was limited on border cells and root surface (Fig. 45). These impairments could be the causal of the reduction of the growth promoting activity in this mutant.

The second mutant which exhibited loss of plant growth activity was the *abrB* gene. Disruption of the *abrB* gene by TnYLB-1 transposon reduced the growth promoting activity in Lemna (26%) as well as in *A. thaliana* (17%) (Fig. 28-31). The *abrB* gene is a DNA-binding global regulator of a plethora of functions that are expressed during the transition from exponential growth to stationary phase and under suboptimal growth conditions (Strauch *et al.* 2005). During this period, the bacterial cells adopt for growth and survival in an altered, nutrient-deprived environment. It is supposed to be one of the most important transition state regulators, because it is involved in the regulation of various cellular functions, such as antibiotic production, competence development, expression of dipeptide transport proteins, proteases, and degradative enzymes, like histidase. *AbrB* displays its regulatory function by acting as a repressor, activator, or preventer (Klein *et al.* 2000). It is unclear how this gene is involved in plant growth stimulation as the pleiotropic *abrB* protein is known to play a role in regulating many antimicrobial metabolites in addition to numerous other genes expressed by postexponential-phase cells (Strauch *et al.* 2007). Hence, we suggest that the loss of growth-promoting activity by *abrB* mutant is more likely due to termination of the metabolic pathway involved in growth stimulation. Through CLSM we observed the limited distribution of the cells when colonizing the root. Most of *abrB* mutant cells resided mostly in border cells of the *A. thaliana* root (Fig. 47) accompanied by defects in biofilm formation within the cells (Fig. 48).

The PGPR RBAM_017410 mutant showed reduction of plant growth promoting activity both in Lemna (42%) and in *A. thaliana* (33%). The complementation of this gene not fully restored the growth promoting ability; however it grew better than control and mutant (Fig. 34-37). The RBAM_017410 is a small peptide with unknown function. Nevertheless, using BLASTX search analyses, the amino sequence of this small peptide shows 76% similarity to ribonucleoside-diphosphate reductase small subunit from *B. subtilis*. In *E. coli*, ribonucleoside diphosphate reductase (NDP reductase) is the only specific enzyme catalyzing the reduction of ribonucleotides to deoxyribonucleotides (dNTPs), the precursors of DNA synthesis, and thus essential to all known life (Guarino *et al.* 2007; and Jiang *et al.* 2008). In the *B. amyloliquefaciens* FZB42 genome, RBAM_017410 gene is located in the region where the gene cluster has still unknown function (Fig. 32). The examination of colonization of this mutant on the root of *A. thaliana* showed no cells appearing on the root (Fig. 49). An alternative explanation for that would be a mishandling during the preparation of the root when monitoring the colonization or may be to the low fluorescence emission from the cells. Using SEM, we found that the cells are impaired in their ability to form biofilm when colonizing the root (Fig. 50). In addition, the shape of the cells was a slim rod which was different form the wild type.

In this research, three PGPR mutants showed the same characteristics that were limited distribution and impairment in biofilm formation when colonizing on the root. Reva et al. 2004 found that the ability for effective initiation of colonization of the root was associated with the ability to form biofilms. The bacterium has to attach to the seed surface, primary roots and root hairs, and to survive from the bacteriocidal substances released by roots. An intriguing finding in this study was the capability of the biofilm mutants to exhibit growth promotion as well as the wild type, even when the colonization on the root was limited. Further investigation is required to reveal this phenomenon. Wang et al. 2005 discovered that quinolinate phosphoribosyltransferase (QAPRTase) was involved in plant growth-promoting activity in *Burkholderia* sp. strain PsJN. However, how QAPRTase participates in the growth promotion signal pathway is still unknown. Another study by Choi et al. 2008 showed the involvement of pyrroloquinoline quinone (PQQ) in plant growth promotion by *P. fluorescens* B16 and suggested that PQQ acts as an antioxidant in plants.

An interesting result which was observed from analyzing the metabolites released by *B. amyloliquefaciens* FZB42 when interacting with *L. minor* in the Steinberg medium was the production of surfactin. This finding confirmed the study of Bais et al. (2004) who demonstrated the role of surfactin from undomesticated *B. subtilis* for biofilm formation and colonization of *A. thaliana* roots. Other lipopeptides and polyketides produced by FZB42 grown in the Landy medium (Chen et al. 2006; Chen et al. 2007; and Koumoutsi et al. 2007), were not detected by MALDI-TOF mass spectrometry performed with the plant extract from Lemna minor and the growth medium, inoculated with FZB42. This implies an important role of surfactin in colonizing plant roots by plant associated *B. amyloliquefaciens* strains. In contrast to bacillomycin D and fengycin, and the polyketides bacillaene, difficidin and macrolactin, which are all produced by FZB42 under laboratory conditions, antimicrobial action of surfactin is relatively weak. However, in *B. subtilis* 6051 the biocontrol ability toward *P. syringae* is related with surfactin formation (Bais *et al.* 2004). In addition, together with fengycin, surfactin of *B. subtilis* strain S449 elicits induced systemic resistance in plants (Ongena *et al.* 2007).

4.4 Identification of genes involved in production of antibiotic and nematocidal

Generating mutants in *B. amyloliquefaciens* FZB42 strain CH5 by employing the transposon TnYLB-1 led to discover two new ribosomally synthesized secondary metabolites named amylocyclicin and plantazolicin. Both antibiotics do not depend on phosphopantetheinyl transferase, like cyclic lipopeptides and polyketides. Amylocyclicin A could be identified as the substance responsible for the reported strong activity of FZB42 against *Bacillus subtilis* HB0042 and also has a strong antibacterial activity against other gram-positive bacteria. This antibiotic belongs to the disparate group I of circular bacteroicins like CclA, lactocyclicin Q, AS-48, circularin A, and uberolysin, with a high pI. On the basis of the predicted secondary structure, amylocyclicin A also contains helical structures und presumably has a gobular structure. It could interact with the membrane of sensitive gram-positive bacteria and form nonselective pores in lipid bilayers, allowing the free diffusion of ions and low molecular weight solutes across the membrane like AS-48 and in a similar way gassericin A and reutericin 6 (Galvez *et al.* 1991; and Kawai *et al.* 2004) leading to cell death or forms anion selective channels in lipid bilayers like CclAI (Gong *et al.* 2009). Amylocyclicin A is the first representative of circular bacteriocins of the genus *Bacillus* belonging to class 2d/IV of bacteriocins (Scholz *et al.* 2011). Plantazolicin is a novel, antibacterial, microcin B17/streptolysin-like compound that was identified in the culture supernatant and cell surface extract from *B. amyloliquefaciens* FZB42. The genetic and biochemical conservation within this particular natural product family has led to a new classification of small, highly modified bacteriocins, the thiazole/oxazole-modified microcins (TOMMs) (Lee *et al.* 2008; and Haft *et al.* 2010).

Another interesting result from the screening based on transposon TnYLB-1 mutants of *B. amyloliquefaciens* FB42 was the finding of four mutants which showed reduction in nematocidal effectiveness (Table. 7). Complementation and characterization of the genes and metabolites involved in nematocidal is still in progress conducted in Yunan University – China.

5. References

Akerley, B. J., Rubin, E. J., Camilli, A., Lampe, D. J., Robertson, H. M., and Mekalanos, J. (1998) Systematic identification of essential genes by *in vitro* mariner mutagenesis. *Proc. Natl. Acad. Sci. USA.* **95**, 8927-8932

Albareda, M., Dardanelli, M. S., Sousa, C., Megias, M., Temprano, F., and Navarro, D. N. R. (2006) Factors affecting the attachment of rhizospheric bacteria to bean and soybean roots. *FEMS Microbiol. Lett.* **259**, 67-73

Ali, N.I., Siddiqui, I.A., Shaukat, S.S., and Zaki, M.J. (2002) Nematicidal activity of some strains of *Pseudomonas* spp. *Soil Biol. Biochemis.* **34**, 1051 – 1058

Allison, C., Emody, L., Coleman, N., and Hughes, C. (1994). The role of swarm-cell differentiation and multicellular migration in the uropathogenicity of *Proteus mirabilis*. *J. Infect Dis.* **69**, 1155-1158

Aloni, R., Aloni, E., Laghans, M., and Ullrich, C.I. (2006) Role of cytokinin and auxin in shaping root architecture: Regulating vascular differentiation, lateral root initiation, root apical dominance and root gravitropism. *Ann. Bot.* **97**, 883-93

Amati, G., Bisicchia, P., and Galizzi, A. (2004) DegU-P represses expression of the motility *fla-che* operon in *Bacillus subtilis*. *J. Bacteriol.* **186**, 6003-6014

Antoun, H., and Kloepper, J. W., 2001 Plant growth-promoting rhizobacteria (PGPR), in: *Encyclopedia of Genetics*, Brenner, S. and Miller, J.H., eds., Academic Press, NY.1447-1480

Ashour, J., and Hondalus M. K. (2003) Phenotypic mutants of the intracellular actinomycete *Rhodococcus equi* created by in vivo *Himar1* transposon mutagenesis. *J. Bacteriol.* **185**, 2644-2652

Atzhorn, R., Crozier, A.. Wheeler, C.T., and Sandberg G. (1988) Production of gibberellins and indole-3-acetic acid by *Rhizobium phaseoli* in relation to nodulation of *Phaseolus vulgaris* roots. *Planta.* **175**, 532-538

Bais, H. P., Park S. W., Weir, T. L., Callaway, R. M. and J. M. Vivanco. (2004) How plants communicate using the underground information superhighway. *Trends Plant Sci.* **9**,26-32

Bais, H.P., Tiffany L.Weir, T.F., Perry, L.G., Gilroy, S., and Vivanco, J.M. (2006) The role of root exudates in rhizosphere interactions with plants and other organisms. *Ann. Rev. Plant Biol.* **57**,233-66

Bano, N., and Musarrat, J. (2004) Characterization of a novel carbofuran degrading Pseudomonas sp. with collateral biocontrol and plant growth promoting potential. *FEMS Microbiol. Lett.* **231**, 13-17

Basset, G.J.S., Quinlivan, E.P., Ravanel,S., Re´ beille, F., Nichols, B.P., Shinozaki, K., Seki, M., Lori C. Adams-Phillips, L. C., Giovannoni, J.J., Gregory, J.F. and Hanson, A.D. (2004) Folate synthesis in plants: The p-aminobenzoate branch is initiated by a bifunctional PabA-PabB protein that is targeted to plastids. *Proc. Natl. Acad. Sci. USA.* **101(6)**, 1496-1501

Bastian, F., Cohen, A., Piccoli, P., Luna, V., Baraldi, R. and Bottini, R.(1998) Production of indole-3-acetic acid and gibberellins A1 and A3 by Acetobacter diazotrophicus and Herbaspirillum seropedicae in chemically defined media. *Plant Growth Regul.* **24**, 7-11

Benhamou, N., Kloepper, J.W., and Hallman, Q.A. Bastian, F., Cohen, A., Piccoli, P., Luna, V. (2002) Induction of defense related ultrastructural modifications in pea root tissues inoculated with endophytic bacteria. *Plant Physiol.* **112**, 919-929

Berg, G., Roskot, N., Steidle, A., Eberl L., Zock, A., and Smalla, K. (2002) Plant-dependent genotypic and phenotypic diversity of antagonistic rhizobacteria isolated from different *Verticillium* host plants. *Appl. Environ. Microbiol.* **68**, 3328-3338

Berg, G., Zachow, C., Lottmann, J., Götz, M., Costa, R., and Smalla, K. (2005) Impact of plant species and site on rhizosphere associated fungi antagonistic to *Verticillium dahliae* Kleb. *Appl. Environ. Microbiol.* **171**, 4203-4213

Berg, G. (2009) Plant-microbe interactions promoting plant growth and health: perspectives for controlled use of microorganisms in agriculture. *Appl. Microbiol. Biotechnol.* **84**, 11-18

Bloemberg, G. V., and Lugtenberg, B. J. J. (2001) Molecular basis of plant growth promotion and biocontrol by rhizobacteria. *Curr. Opin.in Plant Biol.* **4**,343-350

Bolwerk, A., Lagopodi, A. L., Wijfjes, A. H., Lamers, G. E.,. Chin, A. W. T. F., Lugtenberg, B. J., and Bloemberg, G. V. (2003) Interactions in the tomato rhizosphere of two *Pseudomonas* biocontrol strains with the phytopathogenic fungus *Fusarium oxysporum* f. sp. *radicis-lycopersici. Mol. Plant-Microbe Interact.* **16**, 983-993

Bordi, C., Butcher, B.G., Shi, Q., Hachmann, A., Peters, J.E., and Helmann J.D. (2008) In vitro mutagenesis of *Bacillus subtilis* by using a modified Tn7 transposon with an outward-facing inducible promoter. *Appl. Environ. Microbiol.* **74(11)**, 3419-25

Bourhy, P., Louvel, L., Girons, I.S., and Picardeau, M., (2005) Random insertional mutagenesis of *Leptospira interrogans*, the agent of leptospirosis, using a *mariner* transposon. *J. Bacteriol.* **187(9)**, 3255-3258

Branda, S. S., Pastor, J. E. G., Yehuda, S. B., Losick, R., and Kolter, R. (2001) Fruiting body formation by *Bacillus subtilis*. *Proc. Natl. Acad. Sci.* USA. **98(20)**, 11621-11626

Branda, S. S., Pastor, J. E. G., Dervyn, E., Ehrlich, S. D., Losick, R., and Kolter, R. (2004) Genes involved in formation of structured multicellular communities by *Bacillus subtilis*. *J. Bacteriol.* **186(12)**, 3970-3979

Branda, S. S., Vik, A., Friedman, L., and Kolter, R. (2005) Biofilms: the matrix revisited. *Trends Microbiol.* **13(1)**, 20-26

Branda, S. S., Chu, F., Kearns, D. B., Losick, R., and Kolter, R. (2006) A major protein component of the *Bacillus subtilis* biofilm matrix. *Mol. Microbiol.* **59(4)**, 1229-1238

Breton, Y. L., Mohapatra, N. P., and Haldenwang, W. G. (2006) In vivo random mutagenesis of *Bacillus subtilis* by use of tnylb-1, a *mariner*-based transposon. *Appl. Environ. Microbiol*, **72(1)**, 327-333

Chakraborty, U., Chakraborty, B., and Basnet, M. (2005) Plant growth promotion and induction of resistance in *Camellia sinensis* by *Bacillus megaterium*. *J. Basic Microbiol.* **46(3)**, 186-195

Chandler, M., and Mahillon, J. (2002) Insertion sequences revised. In Craig NL, Craigie R, Gellert M, Lambowitz AM, eds. Mobile DNA II. Washington, DC: ASM Press. 305-366

Chen, X.H., Vater, J., Piel, J., Franke, P., Scholz, R., Schneider, K., Koumoutsi, A., Hitzeroth, G., Grammel, N., Strittmatter, A.W., Gottschalk, G., Süssmuth, R.D., and Borriss, R. (2006) Structural and functional characterization of three polyketides synthase gene clusters in *Bacillus amyloliquefaciens* FZB42. *J. Bacteriol.* **188(11)**, 4024-4036

Chen, X. H., Koumoutsi, A., Scholz, R., Eisenreich, A., Schneider, K., Heinemeyer, I., Morgenstern, B., Voss, B., Hess, W. R., Reva, O., Junge, H., Voigt, B., Jungblut, P. R., Vater, J., Süssmuth, R., Liesegang, H., Strittmatter, A., Gottschalk, G., and Borriss, R. (2007) Comparative analysis of the complete genome sequence of the plant growth-promoting bacterium *Bacillus amyloliquefaciens* FZB42. *Nature Biotechnol.* **25(9)**, 1007-1014

Chen, X.H., Koumoutsi, A., Scholz, R., Schneider, K., Vater, J., Süssmuth, R. Piel, J., and Borriss, R. (2009a) Genome analysis of *Bacillus amyloliquefaciens* FZB42 reveals its potential for biocontrol of plant pathogens. *J. Biotechnol.* **140**, 27-37

Chen, X.H., Koumoutsi, A., Scholz, R., and Borriss, R. (2009b) More than anticipated - production of antibiotics and other secondary metabolites by *Bacillus amyloliquefaciens* FZB42. *J. Mol. Microbio. Biotechnol.* **16**, 14-24

Chen, X.H., Scholz, R., Borriss, M., Junge, H., Mögel, G., Kunz, S., and Borriss, R. (2009c) Difficidin and bacilysin produced by plant-associated *Bacillus amyloliquefaciens* are efficient in controlling fire blight disease. *J. Biotechnol.* **140**, 38-44

Chen, X.H. 2009. Whole genome analysis of the plant growth-promoting rhizobacteria *Bacillus amyloliquefaciens* FZB42 with focus on its secondary metabolites. Ph.D. dissertation. Humboldt-Universität zu Berlin, Berlin, Germany.

Choi, O., Kim, J., Kim, J.G., Jeong, Y., Moon, J. S., Park, C. S., and Hwang, I. (2008) Pyrroloquinoline quinone is a plant growth promotion factor produced by *Pseudomonas fluorescens* B16. *Plant Physiol.* **146**, 657-668

Compant, S., Duffy, B., Nowak, J.,Clement, C., and Barka, E. A. (2005) Use of plant growth-promoting bacteria for biocontrol of plant diseases: principles, mechanisms of action, and future prospects. *Appl. Environ. Microbiol.* **71(9)**, 4951-4959

Conrath U, Pieterse C.M.J., and Mauch-Mani, B. (2002) Priming in plant-pathogen interactions. *Trends Plant Sci.* **7**, 210-216

Costerton, J. W., Lewandowski, Z., De Beer, D., Caldwell, D., Korber, D., and James, G. (1994) Biofilms, the customized microniche. *J. Bacteriol.* **176**, 2137-2142

Couillerot, O., Prigent-Combaret, C., Caballero-Mellado, J., Moenne-Loccoz, Y. (2009) Pseudomonas fluorescens and closely-related fluorescent pseudomonads as biocontrol agents of soil-borne phytopathogens. *Lett. Appl. Mirobiol.* **48**, 505-512

Cutting, S.M. and Vander Horn, P.B. (1990a): Molecular Biological Methods for *Bacillus*, Harwood C., Cutting S.M., Eds, Wiley Interscience, Chichister, United Kingdom

Cutting, S.M. and Van der Horn, P.B. (1990b) Genetic Analysis, Harwood C.R., Cutting S.M., Eds, Molecular biological methods for *Bacillus* pp. 27-74, Wiley Interscience, Chichister, United Kingdom

Danielsson, J., Reva, O., and Meijer, J. (2006) Protection of oilseed rape (*Brassica napus*) toward fungal pathogens by strains of plant-associated *Bacillus amyloliquefaciens*. *Microb. Ecol.* **54(1)**, 134-40

Davey, M. E., and O'Toole, G. A. (2000) Microbial biofilm: from ecology to molecular genetics. *Microbiol. Mol. Biol. Rev.* **64(4)**, 847-867

de Salamone, I.E.G., Hynes, R.K., and Nelson, L.M. (2001) Cytokinin production by plant growth promoting rhizobacteria and selected mutants. *Can. J. Microbiol.* **47(5)**, 404-411

Dey R., Pal K.K., Bhatt D.M., and Chauhan, S.M. (2004) Growth promotion and yield enhancement of peanut (*Arachis hypogaea* L.) by application of plant growth-promoting rhizobacteria. *Microbiol. Res.* **159**, 371-394

Dieffenbach, C. W., and Gabriela S. Dveksler, G.S. (1995) PCR Primer: A Laboratory Manual. Cold Spring Harbor Laboratory Press

Dobbelaere, S., and Okon, Y. (**2007**) The plant growth promoting effects and plant responses. In: Elmerich C, Newton WE (eds) Nitrogen fixation: origins, applications and research progress. Associative and endophytic nitrogen-fixing bacteria and cyanobacterial associations, vol V. pp. 145-170, Heidelberg, Springer.

Emmert, E. A. B., and Handelsman, J. (1999) Biocontrol of plant disease: a (Gram) positive perspective. *FEMS Microbiol. Lett.* **171**, 1-9

Flemming, H.C. 1993. Biofilms and environmental protection. *Water Sci. Technol.* **27**, 1-10

Freiberg, C., Fellay, R., Bairoch, A., Broughton, W.J., Rosenthal, A., Perret, X., and Freiberg, C. (1997) Molecular basis of symbiosis between Rhizobium and legumes. *Nature.* **387**, 394-401

Gardener, B. B. MS. (2004) Ecology of *Bacillus* and *Paenibacillus* spp. in agricultural systems. *Phytopathol.* **94**, 1252-1258

Gerhardson, B. 2002. Biological substitutes for pesticides. *Trends Biotechnol.* **20**, 338-343

Glick, B. R., and Bashan, Y. (1997) Genetic manipulation of plant growth-promoting bacteria to enhance biocontrol of phytopathogens. *Biotech. Adv.* **15(2)**, 353-378

Glick, B.R. (2005) Modulation of plant ethylene levels by the bacterial enzyme ACC deaminase. *FEMS Microbiol. Lett.* **251**, 1-7

Glick, B.R., Cheng, Z., Czarny, J., and Duan, J. (2007) Promotion of plant growth by ACC deaminase-producing soil bacteria. *Eur. J. Plant Pathol.* **119**, 329-39

Guarino, E., Jime´nez-Sa´nchez, A., and Guzman, E.C. (2007) Defective ribonucleoside diphosphate reductase impairs replication fork progression in *Escherichia coli*. *J. Bacteriol.* **189(9)**, 3496-3501

Gutierrez-Manero, F.J., Ramos-Solano, B., Probanza, A., Mehouachi, J., Tadeo, F.R., and Talon, M. (2001) The plant-growth-promoting rhizobacteria *Bacillus pumilis* and *Bacillus licheniformis* produce high amounts of physiologically active gibberellins. *Physiol. Plant.* **111**, 206-211

Haas, D., and De´fago, G. (2005) Biological control of soilborne pathogens by fluorescent pseudomonads. *Nat. Rev. Microbiol.* **3**, 307-319

Haft, D. H., Basu, M. K., and Mitchell, D. A. (2010) Expansion of ribosomally produced natural products: a nitrile hydratase- and Nif11-related precursor family. *BMC Biol.* **8**, 70

Haggag, W. M., and Timmusk, S. (2007) Colonization of peanut roots by biofilm-forming *Paenibacillus polymyxa* initiates biocontrol against crown rot disease. *J. Appl. Microbiol.* **104**, 961-969

Hall J.A., Peirson D., Ghosh S., and Glick B.R. (1996) Root elongation in various agronomic crops by the plant growth promoting rhizobacterium *Pseudomonas putida* GR12-2. *Isr. J. Plant Sci.* **44**, 37-42

Halling, S. M., and Kleckner, N. (1982) A symmetrical six-base-pair target site sequence determines Tn*10* insertion specificity. *Cell.* **28**, 155-163

Hamon, M. A., Stanley, N. R., Britton, R. A., Grossman, A. D., and Lazazzera, B. A. (2004) Identification of AbrB-regulated genes involved in biofilm formation by *Bacillus subtilis*. *Mol. Microbiol.* **52 (3)**, 847-860

Hartl, D. L., Lohe, A. R., and Lazovskaya, E. R. (1997) Regulation of the transposable element *mariner*. *Genetica.* **100**, 177-184

Hayes, F. (2003) Transposon - based strategies for microbial functional genomics and proteomics. *Annu. Rev. Genet.* **37**, 3-29

Hontzeas. N., Saleema S., Saleh, S.S., and Glick, B.R. (2004) Induced by acc-deaminase-containing plant-growth-promoting bacteria. *Mol. Plant Microb. Interact.* **17 (8)**, 865-871

Idriss, E. E., Makarewicz, O., Farouk, A., Rosner, K., Greiner, R., Bochow, H., Richter, T., and Borriss, R. (2002) Extracellular phytase activity of *Bacillus amyloliquefaciens* FZB45 contributes to its plant-growth-promoting effect. *Microbiol.* **148**, 2097-2109

Idris, E. E., Iglesias, D. J., Talon, M., and Borriss, R. (2007) Tryptophan-dependent production of indole-3-acetic acid (IAA) affects level of plant growth promotion by *Bacillus amyloliquefaciens* FZB42. *Mol. Plant Microb. Interact.* **20(6)**, 619-626

Inbar, E., Green, S.J., Hadar, Y., and Minz, D. (2005) Competing factors of compost concentration and proximity to root affect the distribution of streptomycetes. *Microbial Ecology.* **50**, 73-81

Jackson, M. B. (1993) Are plant hormones involved in root to shoot communication? *Adv. Bot. Res.* **19**, 104-187

James, T.Y., Boulianne, R.P., Bottoli, A.P.F., Granado, J.D., Aebi, M., And Kües, U. (**2001**) The *pab1* gene of *Coprinus cinereus* encodes a bifunctional protein for *para*-aminobenzoic acid (PABA) synthesis: implications for the evolution of fused PABA synthases, *J. Basic Microbiol.* **42(2)**, 91-103

Jiang, W., Yuna, D., Saleha, L., Bollinger Jr, J.M., and Krebsa, C. (2008) Formation and function of the Manganese(IV)/Iron(III) cofactor in *Chlamydia trachomatis* ribonucleotide reductase. *Biochemistry.* **47(52)**, 13736-13744

Joo, G.J., Kim, Y.M., Lee, I.J., Song, K.S., & Rhee, I.K. (**2004**) Growth promotion of red pepper plug seedlings and the production of gibberellins by *Bacillus cereus, Bacillus macroides* and *Bacillus pumilus*. *Biotechnol. Lett.* **26**, 487-491

Joo, G.J., Kim, Y.M., Kim, J.T., Rhee, I.K., Kim, J.H., and In-Jung Lee, I.J. (2005) Gibberellins-producing rhizobacteria increase endogenous gibberellins content and promote growth of red peppers. *J. Microbiol.* **43(6)**, 510-515

Julien, B., and Fehd, R. (2003) Development of a *mariner*-based transposon for use in *Sorangium cellulosum*. *App. Environ. Microbiol.* **69(10)**, 6299-6301

Kamilova, F., Validov, S., Azarova, T., Mulders, I., and Ben Lugtenberg, B. (2005) Enrichment for enhanced competitive plant root tip colonizers selects for a new class of biocontrol bacteria. *Environ. Microbiol.* **7 (11)**, 1809-1817

Katiyar, V., and Goel, R. (2004) Siderophore mediated plant growth promotion at low temperature by mutant of fluorescent pseudomonad. *Plant Growth Regulation.* **42**, 239-244

Kearns, D. B., Chu, F., Rudner, R., and Losick, R. (2004) Genes governing swarming in *Bacillus subtilis* and evidence for a phase variation mechanism controlling surface motility. *Mol. Microbiol.* **52(2)**, 357-369

Kirov, M.S., Tassell, B.C., Semmler, A.B.T., O'Donovan, L.A., Rabaab, A.A.n and Shaw, J.G. 2002. Lateral flagella and swarming motility in *Aeromonas* Species. *J. Bacteriol.* **184**, 547-555

Klein, K., Winkelmann, D., Hahn, M., Weber, T., and Marahiel, M.A. (2000) Molecular characterization of the transition state regulator AbrB from *Bacillus stearothermophilus*. *Biochimi. Biophys. Acta.* **1493(1-2)**, 82-90

Kloepper, J. W., Ryu, C.M., and Zhang, S. (2004) Induced systematic resistance and promotion of plant growth by *Bacillus spp*. *Phytopathol.* **94**, 1259-1266

Kobayashi, K. (2007) Gradual activation of the response regulator DegU controls serial expression of genes for flagellum formation and biofilm formation in *Bacillus subtilis*. *Mol. Microbiol.* **66(2)**, 395-409

Koumoutsi, A., Chen, X. H., Henne, A., Liesegang, H., Hitzeroth, G., Franke, P., Vater, J., and Borriss, R. (2004) Structural and functional characterization of gene clusters directing nonribosomal synthesis of bioactive cyclic lipopeptides in *Bacillus amyloliquefaciens* Strain FZB42. *J. Bacteriol.* **186(4)**, 1084-1096

Kristich, C. J., Nguyen, V. T., Le, T., Barnes, A. M. T., Grindle, S., and Dunny, G. M. (2008) Development and use of an efficient system for random *mariner* transposon mutagenesis to identify novel genetic determinants of biofilm formation in the core *Enterococcus faecalis* genome. *App. Environ. Microbiol.* **74(11)**, 3377-3386

Kunst, F., and Rapoport, G. (1995) Salt stress is an environmental signal affecting degradative enzyme synthesis in *Bacillus subtilis*, *J. Bacteriol.* **177(9)**, 2403-2407

Lampe, D. J., Akerley, B. J., Rubin, E. J., Mekalanos, J. J., and Robertson, H. M. (1999) Hyperactive transposase mutants of the *Himar1 mariner* transposon. *Proc. Natl. Acad. Sci.. USA.* **96**, 11428-11433

Leach, A.W., and Mumford, J.D. (2008) Pesticide environmental accounting: a method for assessing the external costs of individual pesticide applications. *Environ Pollut.* **151**, 139-47

Lee, S. W., Mitchell, D. A., Markley, A. L., Hensler, M. E., Gonzalez, D., Wohlrab, A., Dorrestein, P. C., Nizet, V., and Dixon, J. E. (2008) Discovery of a widely distributed toxin biosynthetic gene cluster. *Proc. Natl. Acad. Sci.* USA. **105**, 5879–5884

Liu, Z.M., Tucker, A.M., Driskell, L.O., and Wood, D.O. (2007) *Mariner*-based transposon mutagenesis of *Rickettsia prowazekii*. *App. Environ. Microbiol.* **73(20)**, 6644-6649

Lockhart, W.L., Billeck, B.N., and Baron, C. L. (1989) Bioassays with a floating aquatic plant (*Lemna minor*) for effects of sprayed and dissolved glyphosate. *Hydrobiologia.* **188-189**, 353-359

Loper, J. E., and Henkels, M. D. (1999) Utilization of heterologous siderophores enhances levels of iron available to *Pseudomonas putida* in the rhizosphere. *Appl. Environ. Microbiol.* **65**, 5357-5363

Louvel, H., Girons, I. S., and Picardeau, M. (2005) Isolation and characterization of FecA- and FeoB- mediated iron acquisition systems of the spirochete *Leptospira biflexa* by random insertional mutagenesis. *J. Bacteriol.* **187**,3249-3254

Lucy, M., Reed, E., and Glick, B. R. (2004) Applications of free living plant growth-promoting rhizobacteria. *Antonie van Leeuwenhoek.* **86**, 1-25

Lugtenberg, B.J., Dekkers, L., and Bloemberg., G.V. (2001) Molecular determinants of rhizosphere colonization by Pseudomonas. *Annu. Rev. Phytopathol.* **39**, 461-90

Lugtenberg, B. J. J., Chin-A-Woeng, F. C., and Bloemberg, G. V. (2002) Microbe-plant interactions: principles and mechanism. *Antonie van Leeuwenhoek.* **81**, 373-383

Lugtenberg, B., and Kamilova, F. (2009) Plant-growth-promoting rhizobacteria, *Annu. Rev. Microbiol.* **63**, 541-56

Maeder, U., Antelmann, H., Buder, T., Dahl, M. K., Hecker, M., and Homut, G. (2002) *Bacillus subtilis* functional genomics: genome-wide analysis of the DegS-DegU regulon by transcriptomics and proteomics. *Mol. Genet. Genomics.* **268**, 455-467

Maier, T. M., Pechous, R., Casey, M., Zahrt, T. C., and Frank, D. W. (2006) In vivo *Himar1*-based transposon mutagenesis of *Francisella tularensis*. *Appl. Environ. Microbiol.* **72**, 1878-1885

Makarewicz, O., Neubauer, S., Preusse, C., and Borriss, R. (2008) Transition state regulator abrB inhibits transcription of *bacillus amyloliquefaciens* FZB45 phytase through binding at two distinct sites located within the extended *phyC* promoter region. *J. Bacteriol.* **190(19)**, 6467-6474

Malhotra, M., and Srivastava, S. (2009) Stress-responsive indole-3-acetic acid biosynthesis by Azospirillum brasilense SM and its ability to modulate plant growth. *Eur. J. Soil Biol.* **45**, 73 – 80

Marilley, L., and Aragno, M. (1999) Phylogenetic diversity of bacterial communities differing in degree of proximity of *Lolium perenne* and *Trifolium repens* roots. *Appl. Soil Ecol.* **13**, 127-136

Merritt, P.M., Danhorn, T., and Fuqua, C. (2007) Motility and chemotaxis in *Agrobacterium tumefaciens* surface attachment and biofilm formation. *J. Bacteriol.* **189**, 8005-8014

Mireles, J.R., Toguchi, A., and Harshey, R.M. (2001) *Salmonella enterica* serovar typhimurium swarming mutants with altered biofilm-forming abilities: surfactin inhibits biofilm formation. *J. Bacteriol.* **183**, 5848-5854

Moch, C., O. Schrogel, and R. Allmansberger. (1998) The σ^D-dependent transcription of the *ywcG* gene from *Bacillus subtilis* is dependent on an excess of glucose and glutamate. *Mol. Microbiol.* **27**, 889 – 898

Moch, C., Schrogel, O., and Allmansberger, R. (2000) Transcription of the *nfrA-ywcH* operon from *Bacillus subtilis* is specifically induced in response to heat. *J. Bacteriol.* **182**, 4384-4393

Morozova, O.V., Dubytska, L.P., Ivanova, L.B., Moreno, C.X., Bryksin, A.V., Sartakova, M.L., Dobrikova, E.Y., Godfrey, H.P., and Cabello, F.C. (2005) Genetic and physiological characterization of 23S rRNA and *ftsJ* mutants of *Borrelia burgdorferi* isolated by mariner transposition. *Gene.* **357(1)**, *63-72*

Mostertz, J., Scharf, C., Hecker, M., and G. Homuth. 2004. Transcriptome and proteome analysis of *Bacillus subtilis* gene expression in response to superoxide and peroxide stress. *Microbiol.* **150**, 497-512

Müller, H., Westendorf, C., Leitner, E., Chernin, L., Riedel, K., Schmidt, S., Eberl, L., and Berg, G. (2009) Quorum-sensing effects in the antagonistic rhizosphere bacterium *Serratia plymuthica* HRO-C48. *FEMS Microbiol Ecol.* **67**, 468-467

Naumann, B., Eberius, M., and Appenroth, Klaus-J. (2007) Growth rate based dose-response relationships and EC-values of ten heavy metals using the duckweed growth inhibition test (ISO 20079) with *Lemna minor L. clone* St. *J. Plant Physiol.* **164**, 1656-1664

Nielsen T.H., Christopheresen, C., Anthoni, U., and, Sørensen, J. (1999) Viscosinamide, a new cyclic depsipeptide with surfactant and antifungal properties produced by *Pseudomonas fluorescens* DR54. *J. Appl. Microbiol.* **87**, 80-90

Nielsen T.H., Thrane, C., Christophersen, C., Anthoni, U., and Sørensen, J. (2000) Structure, production characteristics and fungal antagonism of tensin — a new antifungal cyclic lipopeptide from *Pseudomonas fluorescens* strain 96. *J. Appl. Microbiol.* **89**, 992-1001

Oliveira, D.F., Carvalho, H.W.P., Nunes, A.S., Silva, G.H., Campos, V.P., Júnior, H.M.S. and Cavalheiro, A.J. (2009) The activity of amino acids produced by

Paenibacillus macerans and from commercial sources against the root-knot nematode *Meloidogyne exigua. Eur. J. Plant Pathol.* **124**, 57-63

Ollinger, J., Song, K. B., Antelmann, H., Hecker, M., and Helmann, J.D. (2006) Role of the *fur* regulon in iron transport in *Bacillus subtilis. J. Bacteriol.* **188(10)**, 3664-3673

Ongena, M., Jourdan, E., Adam, A., Paquot, M., Brans, A., Joris, B., Arpigny, J.L., and Thonart, P. (2007) Surfactin and fengycin lipopeptides of *Bacillus subtilis* as elicitors of induced systemic resistance in plants. *Environ. Microbiol.* **9(4)**, 1084-1090

Ortiz-Castro, R., Valencia-Cantero, E., and Lopez-Bucio, J. (2008) Plant growth promotion by *Bacillus megaterium* involves cytokinin signaling. *Plant Signaling Behavior* **3(4)**, 263-265

Overhage, J., Bains, M., Brazas, M. D., and Hancock, R. E. W. (2008) Swarming of *Pseudomonas aeruginosa* is a complex adaptation leading to increased production of virulence factors and antibiotic resistance. *J. Bacteriol.* **190**, 2671-2679

Perego, M., and Hoch J.A. (1991) Negative regulation of *Bacillus subtilis* sporulation by the *spo0E* gene product. *J. Bacteriol.* **173(8)**, 2514-2520

Persello-Cartieaux, F., Nussaume, L., and Robaglia, C. (2003) Tales from the underground: molecular plant-rhizobacteria interactions. *Plant Cell Environ.* **26**, 189-199

Petit, M. A., Bruand, C., Janniere, L., and Ehrlich. S. D. (1990) Tn*10*-derived transposons active in *Bacillus subtilis. J. Bacteriol.* **172**, 6736-6740

Petzke, L., and Luzhetskyy, A. (2009) In vivo Tn5-based transposon mutagenesis of Streptomycetes. *Appl. Microbiol. Biotechnol.* **83**, 979-986

Peypoux, F., Bonmatin, J.M., and Wallach, J (1999) Recent trends in the biochemistry of surfactin. *Appl Microbiol Biotechnol.* **51**, 553-563

Picardeau, M. (2010) Transposition of fly mariner elements into bacteria as a genetic tool for mutagenesis. *Genetica.* **138**, 551-558

Ping, L., and Boland, W. (2004) Signals from the underground: bacterial volatiles promote growth in Arabidopsis. *Trends Plant Sci.* **9(6)**, 263-6

Raaijmakers, J.M., Vlami, M., and de Souza, J.T. (2002) Antibiotic production by bacterial biocontrol agents. *Antonie van Leeuwenhoek.* **81**, 537-547

Reva, O. N., Dixelius, C., Meijer, J., and Priest, F. G. (2004) Taxonomic characterization and plant colonizing abilities of some bacteria related to *Bacillus amylolyquefaciens* and *Bacillus subtilis*. *FEMS Microbiol. Ecol.* **48(2)**, 249-259

Reznikoff, W.S. (2003) Tn5 as a model for understanding DNA transposition. *Mol. Microbiol.* **47 (5)**, 1199-1206

Rholl, D., Trunck, L., and Schweizer, H.P. (2008) In vivo Himar1 transposon mutagenesis of *Burkhoderia pseudomallei*. *Appl. Environ. Microbiol.* **74**, 7529-7535

Robertson, H. M. (1993) The mariner transposable element is widespread in insect. *Nature.* **362**, 241-245

Rodriguez, H., and Fraga, R., (1999) Phosphate solubilizing bacteria and their role in plant growth promotion. *Biotechnol. Adv.* **17**, 319-339

Rudrappa, T., Quinn, W. J., Wall, N. R. S., Bais, H. P. (2007) A degration product of the salicylic acid pathway triggers oxidative stress resulting in down-regulation of *Bacillus subtilis* biofilm formation on *Arabidopsis thaliana* roots. *Planta.* **226**, 283-297

Ryu, C.M., Farag, M. A., Hu, C.H., Reddy, M.S., Wie, H.X., Pare, P. W., and Kloepper, J. W. (2003) Bacterial volatiles promote growth in *Arabidopsis. Proc. Natl. Acad. Sci.* USA. **100(8)**, 4927-4932

Ryu, C.M., Farag, M. A., Hu, C.H., Reddy, M. S., Kloepper, J. W., and Pare, P. W. (2004) bacterial volatiles induce systemic resistance in Arabidopsis. *Plant Physiology.* **134**, 1017-1026

Sambrook, J., Fritsch, E.F., and Maniatis, T. (1989) Molecular Cloning: a Laboratory Manual, Cold Spring Harbor Laboratory, NY

Scholz, R., Molohon K.J., Nachtigall J., Vater J., Markley A.L., Süssmuth, R.D., Mitchell D.A., and Borriss R. (2011) Plantazolicin, a novel microcin B17/streptolysin S-like natural product from *Bacillus amyloliquefaciens* FZB42. *J. Bacteriol.* **22**, 215-224

Senesi, S., Celandroni, F., Salvetti, S., Beecher, D. J., Wong, A. C. L., and Ghelardi, E. (2002) Swarming motility in *Bacillus cereus* and characterization of a *fliY* mutant impaired in swarm cell differentiation. *Microbiol.* **148**, 1785-1794

Siddiqui, I.A., Haas, D., and Heeb, S (2003) Extracellular protease of *Pseudomonas fluorescens* cha0, a biocontrol factor with activity against the root-knot nematode *Meloidogyne incognita. Appl. Environ. Microbiol.* **71(9)**, 5646-5649

Stanley, N. R., and Lazazzera, B. A. (2005) Defining the genetic differences between wild and domestic strains of *Bacillus subtilis* that affect poly-ß-DL-glutamic acid production and biofilm formation. *Mol. Microbiol.* **57(4)**, 1143-1158

Stanley, N.R., Britton, R.A., Grossman, A.D., and Lazazzera, B. A. (2003) Identification of catabolite repression as a physiological regulator of biofilm formation by *Bacillus subtilis* by use of DNA microarrays. *J. Bacteriol.* **185**, 1951-1957

Steenhoudt, O., and J. Vanderleyden (2000) *Azospirillum*, a free-living nitrogen-fixing bacterium closely associated with grasses: genetic, biochemical and ecological aspects. *FEMS Microbiol. Rev.* **24(4)**, 487-506

Steinberg, F.M., Gershwin, M.E., and Ruckr, R.B. (1994) Dietary pyrroloquinoline quinone: growth and immune response in BALB/C mice. *J. Nutr.* **124**, 744-753

Strauch, M.A., Ballar, P., Rowshan, A.J., and Zoller, K.L. (2005) The DNA-binding specificity of the *Bacillus anthracis* AbrB protein. *Microbiol.* **151**, 1751-1759

Strauch, M.A., Bobay, B.G., Cavanagh, J., Yao, F., Wilson,A., and Breton, Y.L. (2007) Abh and AbrB Control of *Bacillus subtilis* antimicrobial gene expression. *J. Bacteriol.* **189(21)**, 7720-7732

Streker, K., Freiberg, C., Labischinski, H., Hacker, J., and Ohlsen, K., (2005) *Staphylococcus aureus* NfrA (SA0367) is a flavin mononucleotide-dependent NADPH oxidase involved in oxidative stress response. *J. Bacteriol.* **187**, 2249-2256

Timusk,. S. (2003) Mechanism of action of the plant growth promoting bacterium *Paenibacillus polymyxa*. Dissertation. Acta Universitatis Upsaliensis. Uppsala

van Loon, L.C., Bakker P.A.H.M., and Pierterse C.M.J. (1998) Systemic resistance induced by rhizosphere bacteria. *Annu. Rev. Phytopathol.* **36**, 453-483

van Peer, R., Niemann, G.J., and Schippers, B. (1991) Induced resistance and phytoalexin accumulation in biological control of fusarium wilt of carnation by *Pseudomonas* sp. strain WCS417r. *Phytopathol.* **81**, 728-734

van Wees S.C.M., and Glazebrook, J. (2003) Loss of non-host resistance of Arabidopsis *NahG* to *Pseudomonas syringae* pv. *Phaseolicola* is due to degradation product of salicylic acid. *Plant J.* **33**, 733-742

Vassilev, N., Vassileva, M., and Nikolaeva, I. (2006) Simultaneous P-solubilizing and biocontrol activity of microorganisms: potentials and future trends. *Appl Microbiol. Biotechnol.* **71**, 137-144

Verhamme, D. T., Kiley, T. B., and Wall, N. R. S. (2007) DegU co-ordinates multiceluler behaviour exhibited by *Bacillus subtilis*. *Mol. Microbiol.* **65(2)**, 554-568

Verstraeten, N., Braeken, K., Debkumari, B., Fauvart, M., Fransaer, J., Vermant, J., and Michiels, J. (2009) Living on a surface: swarming and biofilm formation. *Trends Microbiol.* **16(10)**, 496-506

Wandersman, C., and Delepelaire, P. (2004) Bacterial iron sources: from siderophores to hemophores. *Annu. Rev. Microbiol.* **58**, 611-647

Wang, K., Conn, K., and Lazarovits, G. (2006) Involvement of quinolinate phosphoribosyl transferase in promotion of potato growth by a *Burkhoderia* Strain. *App. Environ. Microbiol.* **72(1)**, 760-768

Wang, Q., Frye, J.G., McClelland, M., and Harshey, R.M. (2004) Gene expression patterns during swarming in *Salmonella typhimurium* : genes specific to surface growth and putative new motility and pathogenicity genes. *Mol. Microbiol.* **52(1)**, 169-187

Wei, J.Z., Hale, K., Carta, L., Platzer, E., Cynthie Wong, C., Fang, S.C. and Aroian, R.V. (2003) *Bacillus thuringiensis* crystal proteins that target nematodes. *Proc. Natl. Acad. Sci.* USA **100(5)**, 2760-2765

Welbaum, G., A. V. Sturz, Z. Dong, and J. Nowak. (2004) Fertilizing soil microorganisms to improve productivity of agroecosystems. *Crit. Rev. Plant Sci.* **23**, 175-193

Wu, Qingmin., Pei, J., Turse, C., and Ficht, T. A. (2006) Mariner mutagenesis of *Brucella melitensis* reveals genes with previously uncharacterized roles in virulence and survival. *BMC Microbiol.* **6**, 102

Yang, Y., Qi, M., and Mei, C. (2004) Endogenous salicylic acid protects rice plants from oxidative damage caused by aging as well as biotic and abiotic stress. *Plant J.* **40**, 909-919

Yeung, A.T.Y., Torfs, A.C.W., Jamshidi, F., Bains, M., Wiegand, I., Hancock, R.E.W., and Overhage, J. (2009) Swarming of *Pseudomonas aeruginosa* is controlled by a broad spectrum of transcriptional regulators, including MetR. *J. Bacteriol.* **191(18)**, 5592-5602

Youngman, P. J., Perkins, J. B., and Sandman, K. (1985) Use of Tn917-mediated transcriptional gene fusions to *lacZ* and *cat-86* for the identification and study of *spo* genes in *Bacillus subtilis*. *In* J. A. Hoch and P.Setlow (eds.), Molecular Biology of Microbial Differentiation. ASM Press, Washington, D.C. 47-54

Youngman, P. J., Perkins, J. B., and Losick, R. (1983) Genetic transposition and insertional mutagenesis in *Bacillus subtilis* with *Streptococcus faecalis* transposon Tn917. *Proc. Natl. Acad. Sci.* USA. **80**, 2305-2309

Zhang, H., Kim, M.S., Krishnamachari, V., Payton, P., Sun, Y., Grimson, M., Farag, M.A., Ryu, C.M., Allen, R., Melo, I.S., and Pare, P.W. (2007) Rhizobacterial volatile emissions regulate auxin homeostasis and cell expansion in Arabidopsis. *Planta.* **226**, 839-851

Zhang, J. K., Pritchett, M. A., Lampe, D. J., Robertson, H. M., and Metcalf, W. W. (2000) In vivo transposon mutagenesis of the methanogenic archaeon *Methanosarcina acetivorans* C2A using a modified version of the insect *mariner*-family transposable element *Himar1*. *Proc. Natl. Acad. Sci.* USA. **97**, 9665-9670

Acknowledgements

First of all, I would like to thank Prof. Dr. Rainer Borriss for giving me the opportunity to study in his group and also for his guidance, discussion and knowledge that he has shared throughout my study.

DAAD and DIKTI which had given me the financial support that made me possible to study in Germany.

I would also like to thank Dr. Joachim Vater for his mass spectrometric analyses as well as correcting my dissertation and Dr. Wilfrid Bleiss for his scanning electron microscopy analyses.

Next, I would like to thank all of the staffs and students in Bacterial Genetics Department at Humboldt University for the support and nice working atmosphere. Christiana Müller and Sybille Striegl for providing that everything is available for daily activity in the lab. Hua for her initial help. Oliwia for translating the german version of my summary. Kinga, Eva and Lilia for togetherness in the lab and outside activities.

Lastly, I would like to thanks my parents (Bpk. Sukardjo and Ibu Rukmijati) for their endless love and care as well as for being very supportive. My family (my wife Fessy, my daughters Alyssa, Cheryl and Naira) for their support, pray and patient. Lyza and family (Mas Ken and Nesha), Tia and Wulan for their kindness and encouragement to finish my study.

i want morebooks!

Buy your books fast and straightforward online - at one of world's fastest growing online book stores! Environmentally sound due to Print-on-Demand technologies.

Buy your books online at
www.get-morebooks.com

Kaufen Sie Ihre Bücher schnell und unkompliziert online – auf einer der am schnellsten wachsenden Buchhandelsplattformen weltweit! Dank Print-On-Demand umwelt- und ressourcenschonend produziert.

Bücher schneller online kaufen
www.morebooks.de

 VDM Verlagsservicegesellschaft mbH
Heinrich-Böcking-Str. 6-8   Telefon: +49 681 3720 174   info@vdm-vsg.de
D - 66121 Saarbrücken   Telefax: +49 681 3720 1749   www.vdm-vsg.de

Printed by Books on Demand GmbH, Norderstedt / Germany